Introduction to
AV for
Technical Assistants

Albert J. Casciero
and
Raymond G. Roney

LIBRARY SCIENCE TEXT SERIES

Introduction to Public Services for Library Technicians. 3rd ed. By Marty Bloomberg.

Introduction to Technical Services for Library Technicians. 4th ed. By Marty Bloomberg and G. Edward Evans.

Immroth's Guide to the Library of Congress Classification. 3rd ed. By Lois Mai Chan.

Science and Engineering Literature: A Guide to Reference Sources. 3rd ed. By H. Robert Malinowsky, and Jeanne M. Richardson.

The Vertical File and Its Satellites: A Handbook of Acquisition, Processing, and Organization. 2nd ed. By Shirley Miller.

Introduction to United States Public Documents. 2nd ed. By Joe Morehead.

The School Library Media Center. 2nd ed. By Emanuel T. Prostano and Joyce S. Prostano.

The Humanities: A Selective Guide to Information Sources. 2nd ed. By A. Robert Rogers.

Introduction to Library Science: Basic Elements of Library Service. By Jesse H. Shera.

The School Librarian as Educator. By Lillian Biermann Wehmeyer.

Introduction to Cataloging and Classification. 6th ed. By Bohdan S. Wynar, with the assistance of Arlene Taylor Dowell and Jeanne Osborn.

Library Management. 2nd ed. By Robert D. Stueart and John Taylor Eastlick.

An Introduction to Classification and Number Building in Dewey. By Marty Bloomberg and Hans Weber.

Map Librarianship: An Introduction. By Mary Larsgaard.

Micrographics. By William Saffady.

Developing Library Collections. By G. Edward Evans.

Problems in Library Management. By A. J. Anderson.

Introduction to AV for Technical Assistants

Introduction to AV
for
Technical Assistants

By
ALBERT J. CASCIERO
and
RAYMOND G. RONEY

Illustrations by
Bruce Cheeks

LIBRARIES UNLIMITED, Inc.
Littleton, Colorado
1981

The frontispieces for chapters 1, 2, 3, and 4 are reprinted from Sigfred Taubert: Bibliopola. Pictures and Texts about the Book Trade. Two volumes. 4°. Volume I: XXVI, 126 sides with 317 illustrations and 2 facsimiles. Volume II: X, 526 sides with 258 plates, thereof 42 colored, and 2 facsimiles. Cloth DM 36o,--. By permission of Dr. Ernst Hauswedell and Co. Verlag, 2000 Hamburg 13.

LIBRARIES UNLIMITED, INC.
P.O. Box 263
Littleton, Colorado 80160

Library of Congress Cataloging in Publication Data

Casciero, Albert J., 1941-
    Introduction to AV for technical assistants.

    (Library science text series)
    Includes index.
    1. Instructional materials centers. 2. Audio-visual library service. 3. Instructional materials personnel. I. Roney, Raymond G. II. Title. III. Series.
Z674.4.C37   025.3   81-13690
ISBN 0-87287-232-7cl.   AACR2
ISBN 0-87287-281-5pa.

To
Annick and Ruth
Whose patience, encouragement, and
inspiration made this work possible.

PREFACE

More often than not, students entering a program of studies and beginning technical assistants in the field of audiovisual technology are baffled by the extant literature. Numerous textbooks, manuals, and periodicals do provide sound insight to the application of media in the methodology of teaching, and many of these books are good. In fact, many are excellent. But most treat the subject on an advanced level, focusing on highly advanced production techniques or on the cumulative research and its findings. They were written with the professional educator or the specialist in mind and are of little help to the beginner.

This textbook will attempt to provide the aspiring technical assistant with an introduction to the organization, structure, and function of prototype audiovisual centers. It will introduce much of the presently available instructional technology, but, rather than providing extensive and in-depth production coverage, it will guide readers through their first encounters with production techniques. It will also concentrate on the areas that, in the authors' experience, present either a difficult concept or a skill that is truly acquired only after practice. In short, the book provides a shortcut to necessary and relevant information and to hands-on experience and eliminates the need to consult extensive specialized literature or to serve a long period of internship. We hope that after reading this text the technical assistant will be better prepared to undertake the tasks of the profession and be more successful in performing those tasks.

The authors wish to extend their appreciation to Sheila Clark, Sharron Golder, Lucretia Jackson, Doretha Richeson, Maria Willis, Charles Butler, Audrey Jones, Willard Taylor, and Clifton Young, for their assistance toward the completion of this textbook.

Albert J. Casciero

Raymond G. Roney

TABLE OF CONTENTS

PART II
Using Audiovisual Materials and Equipment

PART III
Audiovisual Media in the Communication Process

LIST OF ILLUSTRATIONS

PART I
Managing the Audiovisual Center

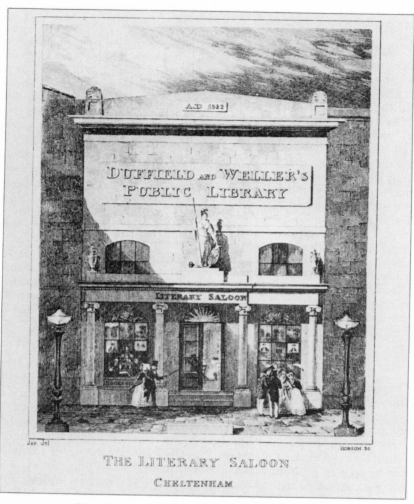

The Literary Saloon, Cheltenham, ca 1850

1–INTRODUCTION TO AV CENTERS

BRIEF HISTORY

Audiovisual or media resources have played an important role in teaching and learning since ancient times. The *camera obscura* was used as a teaching mechanism by Aristotle for his students at the Lyceum in 330 B.C. As early as 1543 A.D., anatomical drawings were used at the University of Padua. In 1870, the magic lantern, a prototype slide projector, was invented and used in Germany.[1] The educational value of motion pictures became evident with the appearance of travelogue shorts accompanying the main feature in the first nickelodeon theaters.

During the early stages of its development, educational media were seen primarily as an aid to didactic instruction. The individualized approach received little consideration; instead, media materials were designed to present a great deal of information to a large group of people. But as technology began to change American society, the library was was no longer a collection of books and electronic communication began to play an integral role in both teaching and learning. In the 1960s Canadian educator, Marshall McLuhan, introduced the idea that "the medium is the message" and stimulated many to believe that contemporary society had moved (or was moving) from a "print" culture to a "visual" one. By the 1970s, however, educators seemed to drift from McLuhanism and toward the realization that the "death of the book" had not occurred and that, if anything, the new media could be used to encourage reading. Thus, the balanced attitude toward the combination of audio and visual messages in the classroom has provided the individual instructor a means of locally producing instructional and learning materials to meet his or her requirements. Along with the implementation of media techniques as part of the teacher training programs, individualized learning strategies have been developing at full speed.

Although books continue to be the more traditional medium, the use of other learning resources has increased significantly. Such media as microforms, motion picture films, disc records, slides, audiotapes, filmstrips, models, art prints, games and simulations, and realia are now circulated. More recently, the use of television, videotapes, and computer terminals has provided added services.

DEFINITIONS

As the newer learning resources have been added, a diversity of names have been developed to identify them. In this book, the term *media* is used in the generic sense to include all types of materials: books, periodicals, films, slides, microform, discs, tapes, etc. The classifications *print* and *nonprint* media have been applied to differentiate the two broad types of materials. Books and periodicals are classified as print media; media requiring the use of special equipment to be seen or heard (e.g., records, filmstrips, tapes, films, etc.) and special objects (e.g., art prints, games, models, etc.) are classified as nonprint. The term *audiovisual* is also used to encompass all nonprint media and special equipment that aid in learning and instruction. The term *AV* or *audiovisual center* is used to define a facility where nonprint media, special equipment, production, and instructional and developmental services are provided in support of instruction and learning. *Library* is used here only to refer to a facility where print media services are provided in support of learning and instruction.

The learning resources that have been in use for years or decades are called the *older media*. Those developed more recently are known as *newer media*. The older media include such materials as slides, films, discs and tapes, filmstrips, transparencies, and sound recordings. The newer media include the more recent developments in electronics, television, tape recorders, and computer applications.

As a reflection of the addition of newer media, the National Education Association (NEA) established a Division of Visual Instruction in 1923 to look at the uses of lantern slides, filmstrips, and silent motion pictures. In 1947, it was changed to the Department of Audio-Visual Instruction so as to include the new media, i.e., sound motion pictures, audio tape recorders, and sound recordings that accompany filmstrips. NEA also established a new Division of Educational Technology in 1970 and the Department of Audio-Visual Instruction was changed to the Association for Educational Communications and Technology (AECT), a separate professional association.[2]

CONCEPT OF AV CENTERS

The technological revolution in education has directed the development of AV centers away from their traditional function as agencies for showing films to that of modern and sophisticated centers disseminating knowledge. Libraries, especially, have felt the impact of the knowledge explosion, and they have responded by providing a variety of services in an attempt to meet the needs of their users in a changing environment. As an aspect of this change, a variety of terms have been used to identify and define the function of these centers. Such terms as media center, instructional resources center, instructional materials center, educational media center, learning resources center, and learning center are used to identify centers in schools, colleges and universities, government agencies, and business and private concerns. Additionally, there seems to be a developing consensus that "learning center" is the more generally applied term.

Today, the learning center is a comprehensive learning facility providing a full range of print and nonprint media, equipment, and services to its users. This is done by creating, producing, and distributing a variety of instructional materials and equipment and by actively participating in the presentation and evaluation of services that can best be utilized to facilitate learning. The concept of the learning center at De Anza College (Cupertino, California, *see* Figure 1-1, page 24) emphasizes the learning center as a learner center where learning is both a product and a process.

Learning centers vary in size and composition. Some are very small facilities with a small number of materials and equipment, as can be found in some schools and small organizations. Others are large, complex multi-service centers like those now found in institutions of higher learning and in large corporations.

The evolution of the learning center has occurred for several reasons. First, the need for lessening the duplication of materials and equipment and for making them more accessible to the users resulted in the merging of the library and audiovisual materials. Second, services (library, audiovisual, instructional development, and nontraditional learning activities) were consolidated into a single administrative unit in order to make them more responsive to the individual learners' needs. Third, in a technological and communication-oriented society, a variety of materials and equipment were required to meet the needs of the users; no one particular format could be appropriate for all.

Figure 1-1. The Learning Center, De Anza College

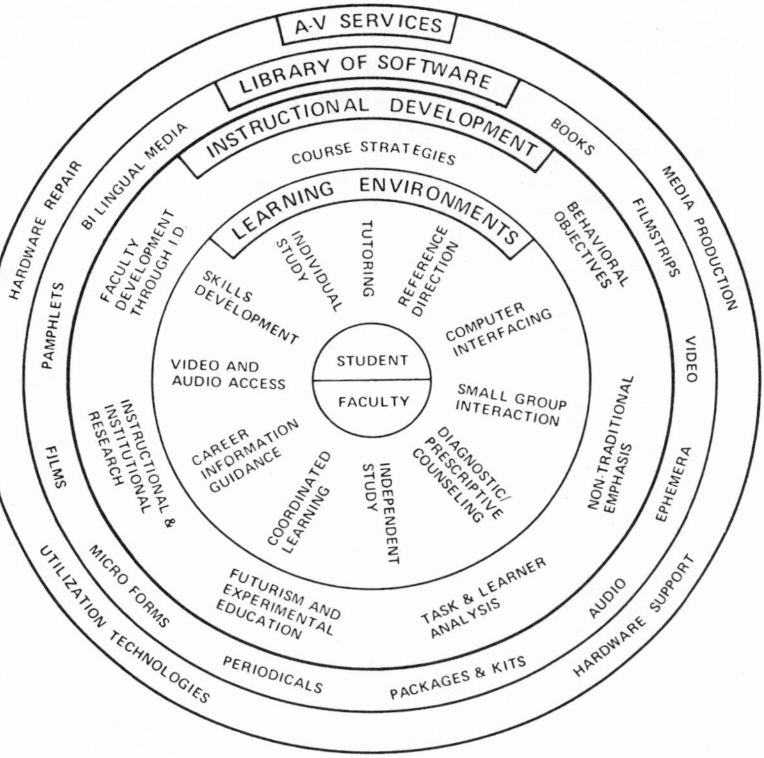

A·V SERVICES

LIBRARY OF SOFTWARE

INSTRUCTIONAL DEVELOPMENT

COURSE STRATEGIES

LEARNING ENVIRONMENTS

BOOKS

BI LINGUAL MEDIA

HARDWARE REPAIR

FACULTY DEVELOPMENT THROUGH I.D.

SKILLS DEVELOPMENT

INDIVIDUAL STUDY

TUTORING

REFERENCE DIRECTION

BEHAVIORAL OBJECTIVES

FILMSTRIPS

MEDIA PRODUCTION

PAMPHLETS

VIDEO AND AUDIO ACCESS

STUDENT

FACULTY

COMPUTER INTERFACING

VIDEO

CAREER INFORMATION GUIDANCE

SMALL GROUP INTERACTION

INSTRUCTIONAL & INSTITUTIONAL RESEARCH

COORDINATED LEARNING

INDEPENDENT STUDY

DIAGNOSTIC/ PRESCRIPTIVE COUNSELING

NON-TRADITIONAL EMPHASIS

EPHEMERA

FILMS

FUTURISM AND EXPERIMENTAL EDUCATION

TASK & LEARNER ANALYSIS

HARDWARE SUPPORT

AUDIO

MICRO FORMS

UTILIZATION TECHNOLOGIES

PERIODICALS

PACKAGES & KITS

Source: Gary T. Peterson, *The Learning Center: A Sphere for Nontraditional Approaches to Education* (Hamden, CT: Linnet Books, 1975), p. 17. Mr. Peterson is vice president of the College of Siskiyous, Weed, California, and has been described as the "father of the learning center movement."

MAJOR CATEGORIES OF AV CENTERS

School Media Centers

As an exponent of innovation, the traditional library has developed into an instructional materials center, media center, or educational media center where a wide range of media materials — print and nonprint — equipment, and services and assistance are available to all its users.

Academic Centers

Community colleges in particular have been the major exponent of change in the college library concept. In an attempt to reflect the growth and function of such centers in higher education, a multiplicity of terms, such as media center, learning resource center, instructional resources center, instructional media center, and learning center, have been applied to these facilities. Community college learning centers have worked to provide a variety of materials and services to fit the more recent thoughts of education, such as individualized learning and making learners more responsible for their learning. These centers also strive to provide a more fluid learning environment that allows learners to proceed at their own pace. Facilities in the community colleges range from small centers offering a few services to large comprehensive, multiservice facilities housed in a multilevel building devoted to this concept.

Special Centers

AV centers (or centers with other names — information center, media center, production center, communication center) are increasingly found in private organizations, industry, and governmental facilities. Such facilities have been designed to meet the special needs of the organization in which they serve through 1) the distribution of materials, 2) the availability of locally created and produced materials, and 3) the accessibility to the patrons of video and audio areas for taping, recording, and listening. The development of such centers in government and industry is rapidly increasing (*see* Figure 1-2).

Figure 1-2. Television Services, Audiovisual Services Department

(Courtesy of Xerox International Center for Training and Management Development)

Public Library Centers

Many public libraries have included nonprint media in their collections. However, the majority of the collections in this area consist of films and recordings. More recently, a number of libraries have included departments for the purpose of graphic and photographic productions, as well as video production.

KINDS OF MATERIALS IN AV CENTERS

The major resources found in the AV center may be classified into four broad categories: visual materials (still and motion pictures); audio materials (disc and tape recording); objects and manipulative materials (realia, games, models, etc.); and Machine-Readable Data Files (magnetic tape, punched cards, aperture cards, etc.).

Since not all centers will have all the materials included within the four categories, only the major forms will be described here.

Visual Materials

Pictures are one of the oldest and most familiar of audiovisual materials. Communication with pictures has been documented since the time of the cave dwellers, and pictures' ability to communicate today is just as potent and valuable.

Still Pictures

Flat opaque pictures (photographic prints and drawings or paintings), slides, filmstrips, transparencies, and microforms constitute several still picture forms. Some of the major advantages of still pictures are: 1) they are versatile in their use with individuals and groups; 2) they stimulate reader interest; 3) they help readers to understand and remember the content of what they are reading; and 4) they are readily available, and the required projection equipment and the pictures can be purchased at low cost.

Specific identification characteristics of these still picture forms are:

1) **Flat Opaque Picture.** A two-dimensional drawing, painting, portrait, photograph, or a print of any of these, produced on an opaque backing (*see* Figure 1-3).

Figure 1-3. Flat Picture

2) Slide. A small unit of transparent material containing an image, mounted in rigid format and designed for use in a slide viewer or projector (*see* Figure 1-4).

Figure 1-4. Slides (with Cassette)

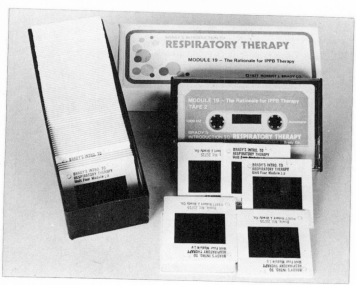

3) Filmstrip. A roll of film containing a succession of images designed to be viewed frame by frame, with or without sound (*see* Figure 1-5).

Figure 1-5. Filmstrip

4) **Transparency.** An image produced on transparent material, designed for use with an overhead projector (*see* Figure 1-6).

Figure 1-6. Transparencies

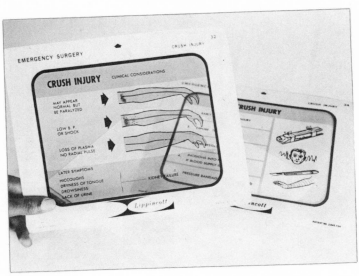

5) **Microform** (microfilm, microfiche, microopaques, aperture cards). A miniature reproduction of printed or other graphic matter which cannot be utilized without magnification (*see* Figure 1-7).

Figure 1-7. Microform Materials (Microfilm and Microfiche)

Motion Pictures

Motion picture forms include 16mm films, 8mm films, and videotapes. Some of the major advantages of motion pictures are: 1) they permit the viewer to visualize a great deal of subject matter of interest; 2) they convey to the viewer an understanding of the subject without requiring solid reading skills; 3) they permit the creativity or re-creativity of events or actions that may or may not occur; and 4) they may be used by the individual, or in small or large groups.

Specific identification characteristics of motion pictures are:

1) **16mm.** This motion picture form has become a standard for educational use. Commonly used in the educational setting for large group viewing, this film may be obtained in color or in black and white with sound (*see* Figure 1-8).

Figure 1-8. 16mm Film

2) **8mm.** The 8mm film has been used for many years, generally for home entertainment. With the recent development of the Super 8mm, this film format has found extensive use in education. A major advantage of this format is the Super 8mm cartridge system, which provides easy handling of the film without the necessity of threading or rewinding. (*see* Figure 1-9, page 30).

Figure 1-9. Super 8mm Film

3) **Videotape.** Produced on tapes 2 inches wide, 1 inch wide, ¾ inch wide, ½ inch wide, or even smaller, this recently developed system for presenting motion has become extremely popular in education. Available both in black and white and in color, the videocassette systems of ½ inch or ¾ inch enable the easy distribution and utilization of motion pictures (*see* Figure 1-10).

Figure 1-10. Videotape and Videocassette

Audio Materials

Audio materials have proven to be extremely effective and valuable resources for learning. The range and quality of recordings continue to improve with time. The major audio formats include disc recordings, cassette tape recordings, reel-to-reel tape recordings, and recorded cards. The value of these resources is that they are flexible in application and are readily available. Also, the equipment necessary for audio materials is easy to operate and portable.

1) **Disc Recordings.** The most common type of educational disc recording is the 33⅓rpm recording (*see* Figure 1-11), although 45rpm and 78rpm recordings are available. The recordings can be played continuously or can easily be stopped and started. They can be used with groups of any size, large or small, and are well suited for individual study.

2) **Tape Recordings.** The older reel-to-reel tape recordings are used less now that the cassette tape recordings have become more popular. The cassette tape is 1/8 inch wide and moves back and forth through a magazine, which encloses the tape, at 1-7/8 inches per second (ips). In addition, they are adaptable and can be used equally as well with large or small groups and for individualized use (*see* Figure 1-11).

Figure 1-11. Disc Recording, Cassette Tape, and Reel-to-Reel Tape

3) **Recorded Cards.** Through the recorded card format, a learner can simultaneously see and hear words or phrases that relate to words, pictures, symbols, or numbers. The card permits one or more learners to listen to the material presented. This medium is especially effective for individualized learning (*see* Figure 1-12, page 32).

Figure 1-12. Recorded Cards with Player

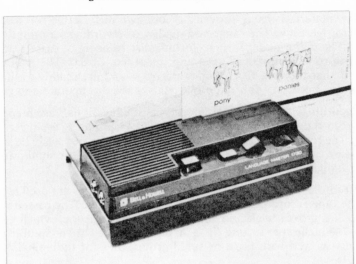

Objects and Manipulative Materials

Objects and manipulative materials have been extremely effective for instructional purposes. Often included in collections located in educational and special centers, these materials permit the patron to manipulate or study the objects, to discover their characteristics, behavior, or operational actions, and to practice with them. The major formats include games, globes, kits, models, realia, etc. The following definitions of these major forms are based on those provided in the glossary of the *Anglo-American Cataloguing Rules*, 2nd edition:[3]

1) **Game.** A set of materials designed for play according to prescribed rules.

2) **Globe.** A sphere with a representation or a map of the earth or of a celestial body.

3) **Kit.** Two or more categories of media which are not fully interdependent and which may be used separately; also designated *multimedia item*.

4) **Model.** A three-dimensional representation of an object, either of the exact size of the original or to scale.

5) **Realia.** Actual objects — artifacts, samples, specimens.

Machine-Readable Data Files

A Machine-Readable Data File (MRDF) comprises a body of information coded by methods that require (typically) a computer for processing. The major forms are files stored on magnetic tape, punched cards (with or without magnetic tape strip), aperture cards, punched paper tapes, disc packs, mark-sensed cards, or optical character recognition font documents. Although these files may not be available in all A V centers, they are extremely important and vital to business and industry and are becoming more important in university and research libraries.

KINDS OF EQUIPMENT IN AV CENTERS

New applications of technology have had a significant impact on the equipment used in instruction. Originally developed for commercial firms, this equipment ranges from the simple to the complex—and from the moderately priced to the very expensive. The variety of equipment available provides the opportunity for any center to offer materials for both group and individualized learning approaches. Again, since not all A V centers will have all the equipment available, only the major forms will be described here.

Filmstrip Projectors and Viewers

Widely used in instruction, filmstrip projectors and 2x2-inch slide projectors have been put to similar use. The filmstrip projector consists mainly of a series of lenses, a lamp, and a reflector; the slide projector is essentially the same except that it uses individual slides. Some projectors can be used for both media. A filmstrip viewer has a built-in screen, which permits greater flexibility in utilization. Many of these projectors and viewers have synchronized audio playback capabilities, which allow the images to be accompanied by sound from audiocassettes or records. These projectors are easy to operate, are quite mobile, and may be used for individual viewing or for use with small or large groups (*see* Figure 1-13 below, and 1-14, page 34).

Figure 1-13. Filmstrip Projector with Remote Control

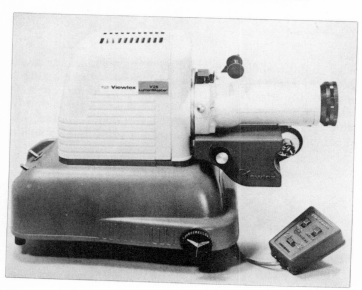

(Courtesy of Viewlex®)

Figure 1-14. Filmstrip Viewer

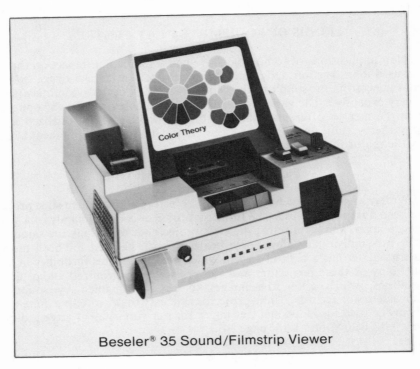

Beseler® 35 Sound/Filmstrip Viewer

(Courtesy of Beseler®)

Slide Projectors

The most common slide projector used in instructional situations is the tray-loading projector. These trays, which are either rotary (*see* Figure 1-15) or rectangular, provide for arrangement of slides in proper order and serve as storage units for slide sets. Recent improvements include automatic slide advance, focus control, remote slide change, and synchronized sound, recording, and playback. Some of these units are self-contained with rear projection capabilities.

Figure 1-15. Slide Projector

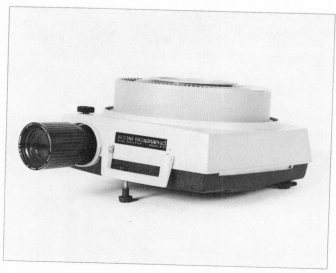

(Courtesy of Kodak®)

Opaque Projectors

Opaque materials (flat materials through which light cannot pass) can be projected with an opaque projector. This is done by means of reflected light. Although a darkened room is usually required because of the lighting deficiency, improvements have been made in the recent projectors which allow effective projection in partially darkened rooms (*see* Figure 1-16).

Figure 1-16. Opaque Projector

Beseler Vu-Lyte® III Opaque Projector

(Courtesy of Beseler®)

Overhead Projectors

The overhead projector (*see* Figure 1-17) sends light through a transparent material and projects it over the head of the operator and onto a screen. This projector is used at the front of the room, which permits the instructor to face the class. The projected image is bright enough for a lighted room.

Figure 1-17. Overhead Projector

Beseler Porta Scribe® G100 Overhead Projector

(Courtesy of Beseler®)

Record Players

The phonograph or record player is used as a playback device for a disc recording. Most can be played at varying speeds: 45, 33⅓, and 16rpm. Longplaying records or LPs recorded at 33⅓rpm consist of 10 to 15 minutes of recorded material per side.

Tape Recorders

Both reel-to-reel and cassette tape recorders are heavily used in AV centers. Relatively inexpensive, tape recorders provide a flexible means of recording and playing back recorded experiences almost anywhere. The audiocard recorder is especially effective for the individualized approach to learning.

Motion Picture Projectors

Although a variety of motion picture projectors are available today for projecting 16mm and 8mm films, all have basically the same system for projecting the images, reproducing the sound, and moving the film through the mechanism (*see* Figure 1-18). Special features include automatic threading, cartridge loading and self-contained screens.

Figure 1-18. 16mm Projector

(Courtesy of Viewlex®)

Television Systems

More recently, television equipment has been utilized and distributed by the AV center. Television receivers and playback units are becoming extremely popular with instructors and students. Videotape in cassettes makes this equipment extremely easy to handle and operate. Receivers are available in both black and white and color.

NOTES

[1]Irving R. Merrill and Harold A. Drob, *Criteria for Planning the College and University Learning Resources Center* (Washington, DC: Association for Educational Communications and Technology, 1977), p. 1.

[2]Merrill and Drob, *Criteria*, p. 3.

[3]Michael Gorman and Paul W. Winkler, eds., *Anglo-American Cataloguing Rules*, 2nd ed. (Chicago: American Library Association, 1978).

BASIC SOURCES

Brown, James W., Kenneth D. Norberg, and Sara K. Suygley. *Administering Media: Instructional Technology and Library Services*, 2nd ed. New York: McGraw-Hill, 1972.

Brown, James W., Richard B. Lewis, and Fred F. Harcleroad. *A V Instruction*, 5th ed. New York: McGraw-Hill, 1977.

Burlingame, Dwight F., Dennis C. Fields, and Anthony C. Schulzetenberg. *The College Learning Resource Center*. Littleton, CO: Libraries Unlimited, 1978.

Cabeceiras, James. *The Multimedia Library: Materials Selection and Use*. New York: Academic Press, 1978.

Merril, Irving R., and Harold A. Drob. *Criteria for Planning the College and University Learning Resources Center*. Washington, DC: Association for Educational Communications and Technology, 1977.

Peterson, Gary T. *The Learning Center: A Sphere for Nontraditional Approaches to Education*. Hamden, CT: Linnet Books, 1975.

Schroeder, Don. *Audiovisual Equipment and Materials*. Metuchen, NJ: Scarecrow Press, 1979.

Veit, Fritz. *The Community College Library*. Westport, CT: Greenwood Press, 1975.

The Library of Lackington Allen & Co., Finsbury Square, London, ca 1795

The Library of Lackington Allen & Co., Finsbury Square, London, ca 1795

2 – ORGANIZATION AND ADMINISTRATION OF AUDIOVISUAL SERVICES

The center's organizational structure will vary according to the objectives of the organization it serves. Generally, two basic organizational patterns are most often used today: the audiovisual center and the learning center.

THE AUDIOVISUAL CENTER

The audiovisual center operates as an independent unit within an organization, business, industry, or specialized concern. Commonly found in federal government agencies, large corporations, and associations, these centers support informational, training, and promotional efforts by providing all audiovisual services, exclusive of print materials, for the parent organization. They vary from very small centers with limited capabilities to multifaceted centers that provide a full range of media support services, but usually, a separate library is maintained within the same organization for the utilization and distribution of print materials. Two examples of the AV centers are shown in Figures 2-1, page 40, and 2-2 on page 41.

Figure 2-1. Organization Chart
U.S. Department of Housing and Urban Development, Office of Public Affairs

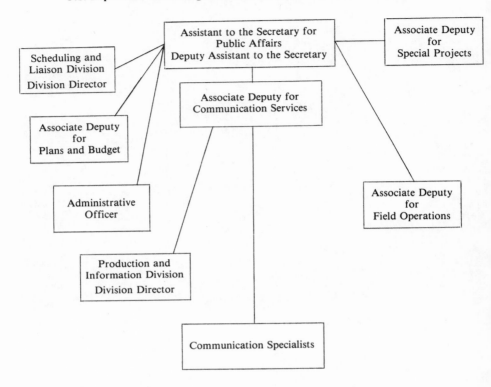

In this organization chart (Figure 2-1), the Production and Information Division provides audiovisual services to a specialized unit within the larger organization. The division director is directly responsible to the Associate Deputy for Communication Services. Within the Xerox Corporation (Figure 2-2), the AV center has the responsibility for providing instructional and production services for the entire organization. The manager reports directly to the Manager of Operations and Administration.

Figure 2-2. Organization Chart
Xerox Corporation, International Center for Training and Management Development

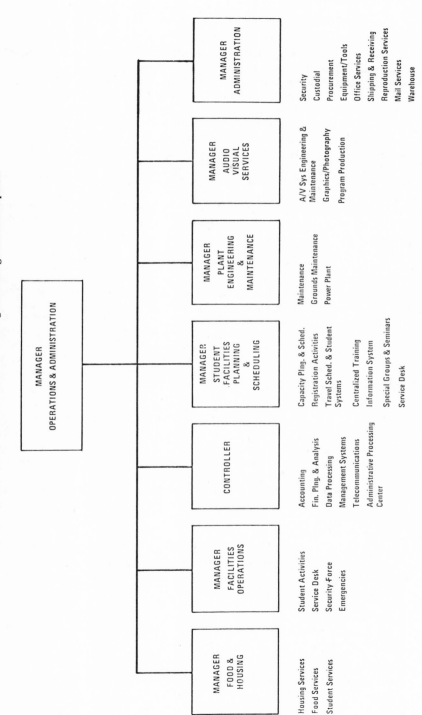

(Courtesy of Xerox International Center for Training and Management Development)

THE LEARNING CENTER

The learning center has merged library and audiovisual facilities into a single comprehensive unit. Utilizing what is often referred to as a centralized organizational approach, the learning center places both print (library) and nonprint (audiovisual) services under the responsibility of a single director or dean. This approach is most often found in educational systems and institutions. School systems and two-year colleges have been pioneers in this direction. In fact, for more than a decade now, school systems have adopted the multimedia concept for the organization and provision of media services, and, more recently four-year colleges and universities have been adopting this pattern also. Although their names vary—media center, multimedia center, resource center, educational material center, etc.—these centers are considered broadly as "learning centers." Figure 2-3 illustrates this organizational pattern.

As indicated in this organizational chart, an Associate Director for Library Services and an Associate Director for Media Services report to the Deputy Director of Library and Media Services.

TYPES OF SERVICE

Although the organizational structure may vary according to the dictates and purposes of the parent organization, many services are provided by most AV centers (production services may be limited or not included in some centers)—specifically, 1) distribution and utilization, and 2) production services. Within these two major areas, the following activities and operations are provided:

1) Distribution and Utilization

— Evaluation of materials and equipment
— Selection and procurement of materials and equipment
— Organization and maintenance of materials and equipment
— Circulation and distribution of materials and equipment
— Instruction in the operation and utilization of equipment and materials
— Promotion and use of AV materials and services
— Consultation with users and provision of information for reference queries

2) Production Services

Graphics:
— Preparation of visual materials for use by both faculty and administration. These include transparencies, charts, posters, lettering, graphs, diagrams, illustrations, dry mounting, laminating, visuals for motion pictures and television.

Still Photography:
— Production of photographs through the photographing and processing of prints and slides

Motion Picture Photography:
— Production of instructional film including both 16mm and Super 8mm films

Figure 2-3. Library and Media Services Division Organizational Structure University of the District of Columbia

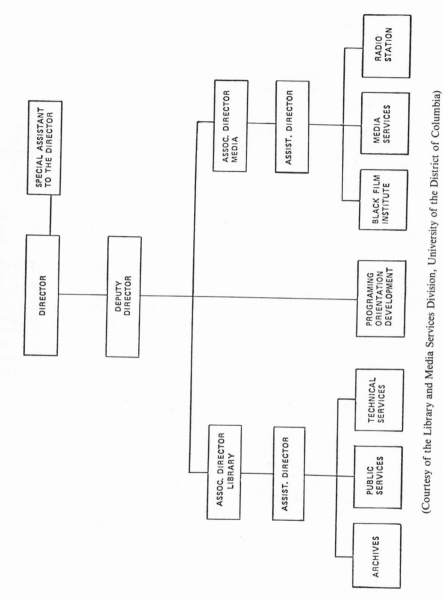

(Courtesy of the Library and Media Services Division, University of the District of Columbia)

Television:
— Production of videotape recordings
— Provision of studio and closed circuit television facilities

Audio:
— Production of audiotape recordings
— Duplication of audiotapes
— Production and provision of facilities for radio programs

Although these services are provided for all users, in some instances various kinds of special facilities, such as a reading lab, a language lab, a testing lab, or a translations unit (in technical centers), are provided according to the purposes of the organization. Further, in order to assist teachers or faculty members to develop and utilize instructional materials and equipment, a faculty resource unit may be a part of the center. An example of such a unit at one university is as follows:

Faculty Resource Unit (University of the District of Columbia)

The purpose of this unit is to provide faculty with the resources to support their learning activities. To accomplish this, there are five major operational components that have clearly identified objectives.

1) *Selection.* Faculty will be assisted with the identification and selection of learning materials. This is to be accomplished through the maintenance of a *Catalog* collection. These catalogs should provide bibliographic information on media materials. The collection is to include catalogs obtained from distributors and production companies that deal with media. *In-house developed catalogs* and bibliographic listings will augment and support commercially available reference tools.
Handbooks relating to preparation and orientation in the use of media materials should be also collected and developed. The development and upkeeping of an up-to-date *collection catalog* will be the responsibility of this component. In addition, the *dissemination* of information about new materials will originate here. This can be accomplished by leaflets, materials obtained on approvals, displays, etc.

2) *Obtaining Materials.* The unit will maintain an efficient *rental service* of media materials such as films, videotapes, multimedia kits, etc. for faculty preview and classroom presentation.

3) *Methods.* Media specialists will provide advice on *strategies* and *methodology* in the application of media in learning. Faculty will be provided with *consultation* on innovative media approaches to learning. Additionally, the identification of media methodology will be supported by *designing* appropriate instructional strategies. Specifying available materials and producing other materials for the faculty to integrate in the identified methodology is also the logical complement to the design function and will be the responsibility of this unit.

4) *Preview.* The facilities will provide an environment with the required equipment for *individual* or *small group viewing* of all media. Provisions for storage of classroom materials identified for faculty use on an ongoing basis is also necessary. Faculty workshops in the *utilization of materials* and *operation of equipment* will be offered regularly. Opportunity for hands-on practice in the handling of media hardware and software will be scheduled frequently and also given on an individual basis as requested.

TYPES OF PERSONNEL

The personnel of the AV center must have broad educational preparation in the area of media and must possess the necessary competencies in a variety of areas within communications. It is necessary that all personnel have a basic understanding of both print and nonprint media as well as specific skills necessary for the performance of duties. The professional and support staff consists of media specialists, media technical assistants or technicians, and media aides.

Professional Staff

The professional staff includes those persons who possess advanced degrees, experience in media or related areas, and competence in planning and administering a media program.

The *media specialist* possesses a master's degree in either media, library science, educational communication, or instructional technology and should have competencies in the areas of media utilization and operation, learning, curriculum, and supervision. Media specialists with specialization in fields other than media are also included as part of the professional staff. Some of these fields are photography, English, business, art, engineering, psychology, and education. Very often, specialization of this type will be an asset for aspiring specialists who, for instance, intend to become scriptwriters, who are interested in still and motion picture production, or who are interested in the management and design of instructional facilities and systems.

The following job description lists the duties and responsibility a media specialist is expected to carry out.

Media Specialist/Position Control:
Under the general supervision of the Director, Dean or Coordinator, the Media Specialist assists in the development and effective operation of the Center.

Duties and Responsibilities:
— Promotes, develops and suggests improved methods of instructional support among faculty and staff through communication and greater utilization of equipment and resources
— Assists the administration and its organization of AV services
— Coordinates and supervises the acquisition and maintenance of all audiovisual equipment and supplies
— Coordinates and supervises the scheduling and use of the AV equipment
— Evaluates and selects AV materials
— Coordinates and supervises the preparation and effective utilization of instructional materials
— Promotes and utilizes all types of AV resources and equipment
— Supervises and evaluates personnel in the unit
— Maintains an up-to-date inventory of all pieces of AV equipment
— Keeps abreast of new equipment, materials, and production techniques in the AV field and acquaints faculty with latest developments
— Assists in budget preparation

Support Staff

The support staff includes both the media technical assistant or technician and the media aide. They work with, and under the direction of media specialists, who plan and direct the operations of the AV centers and libraries.

In 1977, the Association for Educational Communications and Technology (AECT) published the *Guidelines for Certification and Media Specialists*, which presented an in-depth report and refinement of all National Certification Task Force documents, the AECT Certification Model originally issued in 1974. The guidelines state the need for certification of educational communications and technology personnel employed in educational systems and institutions. The guidelines, which are competency based, define three levels of competency: 1) entry or aide positions, 2) middle or technician positions, and 3) advanced or specialist positions. The following clarification of definitions from the *Jobs in Instructional Media* (*JIMS*) is also included in the guidelines.[1]

Entry (Aide)

Entry level personnel are given *specific* instructions about the tasks they perform. The tasks may be only part of a process, the other parts of which the worker cannot or does not control. Entry level personnel can be trained for a task in a relatively short period of time, since almost everything they need to know is contained in the task. They are not required to solve problems external to the task. If something happens which is not covered by the instructions, the entry level worker asks for help and cannot be held responsible for solving the problem.

Middle (Technician)

Instructions given to middle level personnel deal more with a cluster of tasks leading to a *specified output* or outcome. These staff members have a broader view of situations and are expected to generalize more from task to task than are personnel at the entry level. The middle level worker is responsible for the product as long as all of the routines necessary to reach the output have been made specifically available.

Media Technical Assistant or Technician

The *Library Education and Manpower* policy (which was reissued in 1976 under the title, *Library Education and Personnel Utilization*) was adopted by the Council of the American Library Association in 1970. Commenting on the *technical assistant*, the statement indicated that this category assumes

> certain kinds of specific 'technical' skills; they are not meant simply to accommodate advanced clerks. While clerical skills might well be part of a Technical Assistant's equipment, the emphasis in an assignment should be on the special technical skill.[2]

The policy also includes the following description:[3]

Figure 2-4. Categories of Library Personnel — Supportive

Title	Basic Requirements	Nature of Responsibility
Library Technical Assistant Technical Assistant	At least two years of college level study OR A.A. degree with or without Library Technical Assistant training; OR post-secondary school training in relevant skills.	Tasks performed as supportive staff to Associates and higher ranks, following established rules and procedures, and including, at the top level, supervision of such tasks.

Based upon this statement of policy, the *Criteria for Programs to Prepare Library/Media Technical Assistants* was adopted by the Library Education Division of the American Library Association as the official policy of the division. Intended to serve as a guide for planning programs or for the evaluation of existing programs for Library/Media Technical Assistants, these criteria define the nature of the work of the LMTA as follows:

1. The LMTA performs supportive paraprofessional tasks under the direction of a librarian or supervisor. As a member of the personnel team he carries out operations and services essential to effective functioning of the organization.

2. The work of the LMTA may fall within one or more functional areas of library or center operations, for example technical processes, or audiovisual materials.[4]

AECT's 1977 *Guidelines for Certification of Media Specialists* identified the three major areas of responsibility of personnel in educational communications and technology: 1) media management, 2) media product development, and 3) instructional program development. Within each area of responsibility two levels of complexity — technician and specialist — were also identified. Nine basic media functions were indicated within each major area of responsibility, and on each level of complexity:

1) Organization management
2) Personnel management
3) Research and theory
4) Design
5) Production
6) Evaluation and selection
7) Support and supply
8) Utilization
9) Utilization/dissemination

The following competencies were listed under each specialization:

Competencies for Technicians in Media Management

Organization Management Outcomes
- Keeping of purchase records/accounts
- Keeping of work records/payment records
- Keeping of student records
- Keeping of miscellaneous records
- Billing of clients
- Filing of materials
- Mailing/shipping of materials
- Typing
- Performance of minor clerical activities
- Ordering of materials
- Purchasing of materials
- Analyses of organization components
- Scheduling meetings/appointments
- Determining finances/financial constraints
- Seeking funds
- Computing budgets/financial records
- Approving/evaluating of work/products
- Determining need for equipment/facilities/personnel/ procedures
- Making available equipment/facilities/personnel
- Checking for accuracy
- Determining and enforcing time constraints/deadlines
- Writing time/pert charts
- Selection of personnel/materials/equipment and procedures for management
- Administration/coordination of projects
- Monitoring/changing of organization structure and goals
- Writing of work plans/management reports
- Clarification of management goals

Personnel Management Outcomes
- Hiring of staff
- Evaluation/assessment of personnel output
- Assignment of work to personnel/outside consultants
- Assistance in communications between management and staff
- Interaction with individual and groups

Research—Theory Outcomes
- Collation/summarizing of data in preparation for analysis
- Analysis of data
- Interpretation of data

Evaluation—Selection Outcomes
- Diagnoses of equipment defects
- Assessment of materials
- Assessment of devices

Support Supply Outcomes
- Maintenance of equipment/materials
- Picking up and delivery of equipment and materials
- Repair of equipment
- Keeping of repair records
- Repair and inspection of materials
- Keeping of equipment inventory
- Labeling of equipment and materials
- Storage of equipment and materials
- Verification of orders/lists
- Compiling of files/checklists
- Transmission of videotapes
- Distribution/circulation of materials
- Scheduling of materials/equipment/facilities
- Cataloging of materials
- Operation of computer terminals
- Installation of equipment
- Preparation for multimedia presentations
- Ordering of films/materials/equipment
- Location of materials
- Ordering of replacement materials
- Transmission of radio broadcasts

Utilization Outcomes
- Preparation for learning activities
- Presentation of information

Utilization — Dissemination Outcomes
- Distribution of information
- Discussion (two way interaction)
- Teaching (formal interaction)

Competencies for Technicians in Media Product Development

Organization Management Outcomes
- Assignment/coordination of work of units/personnel
- Determining and enforcing time constraints/deadlines

Personnel Management Outcomes
- Hiring of staff

Production Outcomes
- Production of overhead transparencies
- Production of photographic materials
- Production of printed materials
- Production of audio recordings
- Production of TV recordings
- Production of CAI materials
- Production of motion pictures
- Production of contour map
- Getting approval of materials
- Design of art work/layout
- Production of prototype devices

- Production of multiple copies/prototype materials
- Organization of components

Evaluation—Selection Outcomes
- Monitoring of equipment operation
- Verification of instructional system components

Competencies for Technicians in Instructional Program Development

Personnel Management Outcomes
- Interaction with individuals and groups

Research—Theory Outcomes
- Analysis of data
- Interpretation of data

Design Outcomes
- Analysis and description of learners
- Analysis and description of content
- Design of pre and post tests

Evaluation—Selection Outcomes
- Assessment of people
- Assessment of materials
- Assessment of techniques

Utilization—Dissemination Outcomes
- Distribution of information
- Discussion (two way interaction)
- Teaching (formal interaction)
- Explanation (informal interaction)

CAREER CATEGORIES

The relationship of the technical assistant to other personnel within an audiovisual center was commented upon and illustrated in the *Library Education and Personnel Utilization: Policy*.[2]

If one thinks of Career *Lattices* rather than Career *Ladders*, the flexibility intended by the Policy Statement may be better visualized. The movement among staff responsibilities, for example, is not necessarily directly up, but often may be lateral to increased responsibilities of equal importance. Each category embodies a number of promotional steps within it, as indicated by the gradation markings on each bar. The top of any category overlaps in responsibility and salary the next higher category.[5]

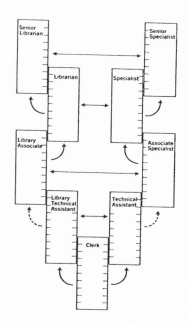

The following "Comments on the Categories" are reprinted by permission of the American Library Association from *Library Education and Personnel Utilization*, pp. 4-5.

The *Clerk* classifications do not require formal academic training in library subjects. The assignments in these categories are based upon general clerical and secretarial proficiencies. Familiarity with basic library terminology and routines necessary to adapt clerical skills to the library's needs is best learned on the job.

The *Technical Assistant* categories assume certain kinds of specific "technical" skills; they are not meant simply to accommodate advanced clerks. While clerical skills might well be part of a Technical Assistant's equipment, the emphasis in an assignment should be on the special technical skill. For example, someone who is skilled in handling audiovisual equipment, or at introductory data processing, or in making posters and other displays might well be hired in the Technical Assistant category for these skills, related to librarianship only to the extent that they are employed in a library....

The *Associate* categories assume a need for an educational background like that represented by a bachelor's degree from a good four-year institution of higher education in the United States. Assignments may be such that library knowledge is less important than general education, and whether the title is *Library* Associate or Associate *Specialist* depends upon the nature of the tasks and responsibilities assigned. Persons holding the B.A. degree, with or without a library science minor or practical experience in libraries, are eligible for employment in this category. The titles within the

Associate category that are assigned to individuals will depend upon the relevance of their training and background to their specific assignments.

The Associate category also provides the opportunity for persons of promise and exceptional talent to begin library employment below the level of professional (as defined in this statement) and thus to combine employment in a library with course work at the graduate level. Where this kind of work/study arrangement is made, the combination of work and formal study should provide 1) increasing responsibility within the Associate ranks as the individual moves through the academic program, and 2) eligibility for promotion, upon completion of the master's degree, to positions of professional responsibility and attendant reclassification to the professional category.

The first professional category — *Librarian,* or *Specialist* — assumes responsibilities that are professional in the sense described [above]. A good liberal education plus graduate-level study in the field of specialization (either in librarianship or in a relevant field) are seen as the minimum preparation for the kinds of assignments implied. The title, however, is given for a position entailing professional responsibilities and not automatically upon achievement of the academic degree.

The *Senior* categories assume relevant professional experience as well as qualifications beyond those required for admission to the first professional ranks. Normally it is assumed that such advanced qualifications shall be held in some speciality, either in a particular aspect of librarianship or some relevant subject field. Subject specializations are as applicable in the *Senior Librarian* category as they are in the *Senior Specialist* category.

Working under guidance and supervision, technical assistants play an important role in carrying out the operations and services of an audiovisual center. In so doing, they provide valuable support and assistance to the specialists, freeing them to perform important professional activities.

The technical assistant, who may be employed in a variety of audiovisual facilities, should have a liberal education background. This should include a broad base of general education courses in the physical sciences, mathematics, social sciences, humanities, and communications. Such a background will equip the technician with the ability to communicate effectively and to understand methods and procedures in audiovisual facilities in both educational — elementary, secondary, higher education — and specialized organizations, both private and government.

An associate degree is generally considered to be the major entry requirement to the technical assistant category in libraries and audiovisual centers. In *New Careers Project Lattice*, Gerald J. Moffitt outlines a library/media career lattice with entry requirements (*see* Figure 2-5).

Figure 2-5. Library/Media Center Career Lattice

Library/Media Assistant I	Library/Media Technician II
Library/Media Assistant II	Library/Media Technician III
Library/Media Technician I	Library/Media Specialist

Entry-Level Requirements

Library/Media Assistant I:

High school graduate or G.E.D., enrolled in the Library Technical Assistant program. No experience.

Library/Media Assistant II:

Successful completion of one year in the Library Technical Assistant program and continuing satisfactory work performance and development during the previous year.

Library/Media Technician I:

Associate degree in Library Technical Assistant program or at least a third-year student in the College of Education majoring in library education with two (2) years of experience in library/media work.

Library/Media Technician II:

B.Ed. in library education and one (1) year of experience in library/media work; or associate in library technical assistant program with a minimum of two (2) years full-time experience in library/media work following the awarding of the degree; or, equivalent in a relevant special field.

Library/Media Technician III:

B.Ed. in library education and an equivalent to three (3) years full-time experience in library/media work; or associate in library technical assistant and five (5) years experience in library/media work of which the last two (2) years must have been at the Technician II level; or, equivalent in a special relevant field.

Librarian/Media Specialist:

Master's degree in library science or media education, and two (2) years of library media work experience.

(Reprinted with permission from Gerald J. Moffitt, *New Careers Project Lattice* [Toledo, OH: University of Toledo, 1970.])

PROFESSIONAL ASSOCIATIONS AND TRAINING PROGRAMS

Professional associations and training programs are extremely important to the continued growth and development of library and media technical assistants.

Through training programs and association meetings and conferences, they provide an opportunity for technical assistants to acquire new concepts and techniques in the field. In fact, in order to qualify for a variety of jobs in the AV or learning center, formal training beyond high school is often essential. Many training programs are offered by two-year institutions and other institutions of higher learning. Generally, these programs consist of either media/audiovisual courses or a combination of library and media courses including general education courses relevant to library/media work that are designed to be completed within a two-year period. A directory entitled *Directory of Institutions Offering or Planning for the Training of Library-Media Technical Assistants* is published fairly frequently by the Council on Library/Media Technical-Assistants (COLT). The directory lists by geographical location the many library-media programs offered at higher educational institutions around the country. (The forthcoming 1981 directory can be obtained by writing to Suzanne Gill, Publication Chairperson, COLT, 11920 Hargrove, Des Peres, MO 63131.)

Since 1966, COLT, an international association of people and organizations interested in the training and employment of library-media support staff, has promoted and contributed to the improved status of technical assistants. COLT has provided not only a meeting ground for technical assistants across the nation to develop camaraderie but also support for their mutual goals. Its membership is open to any person or organization. (Application for membership and requests for copies of publications should be sent to Loretta K. Harris, Library of the Health Sciences, P.O. Box 7509, Chicago, IL 60680.)

The two organizations that have done the most for the audiovisual and the library professions are the Association for Educational Communications and Technology (AECT) and the American Library Association (ALA). The AECT is the professional association for audiovisual and instructional materials specialists, educational technologists, audiovisual and television production personnel, and teacher educators. The association promotes the improvement of education through the systematic planning, application, and production of communications media for instruction. Although the present edition is outdated, AECT has in the past published a directory of *Planning Programs for Educational Media Technicians*. (Requests for additional information regarding the association's activities should be sent to the Association for Educational Communications and Technology, 1126 16th St., NW, Washington, DC 20036.)

ALA is the professional association for librarians and others interested in librarianship. The association was established to promote libraries and librarianship and to establish standards in libraries of all kinds. The Subcommittee on the Training of Library Supportive Staff was established through the Standing Council on Library Education to analyze educational needs and to develop criteria for education and training programs for the different categories of support staff. The subcommittee is responsible for reporting to ALA significant developments in this area and for suggesting action or research related to support staff. (Requests for additional information may be obtained by writing to the American Library Association, Office for Library Personnel Resources, 50 East Huron St., Chicago, IL 60611.)

FACILITIES

Adequate media facilities are of paramount importance to the support of the center's activities and programs. The facilities should be designed to provide effective service to the total program. Although decisions regarding the necessary areas of the center, and the size of and relationships among these areas vary according to the needs of the parent organization and its program activities, all areas should be interrelated to promote the effective operation of the audiovisual services. Within the learning center complex, the audiovisual facilities may be separate or completely integrated within one facility. Of major importance to such a facility is the provision of opportunities for both group and individual learning. Locations where the individual user is able to work with books, films, slides, videotapes, audio recordings, etc. should be provided. Similarly, group learning areas should be provided so that all group activities have the optimum viewing and listening conditions. Figure 2-6 shows large group learning areas, as well as individual and seminar learning areas.

Figure 2-6. Group Learning Areas

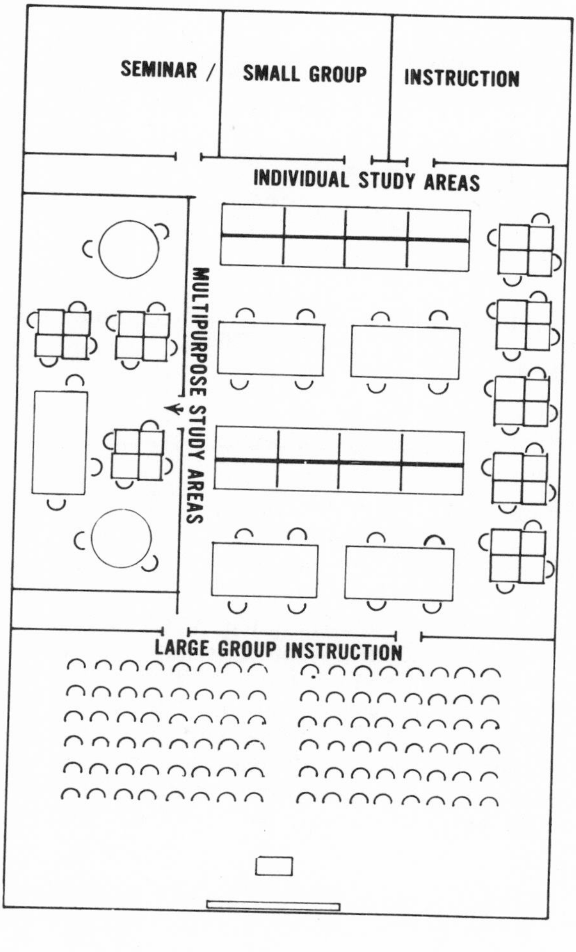

Production and storage areas included within the facility should be carefully planned and designed in order that materials can be produced and used effectively. Production areas may include one or all of the following: graphic, photographic, video and audio (*see* Figure 2-7). Some of these areas will require special utilities, such as necessary and strategically placed electrical outlets, running hot and cold water, special lighting for photographic dark rooms, and high intensity lighting for TV production and indoor filming. These requirements may increase according to the degree of complexity of the facilities.

Figure 2-7. Production and Storage Areas

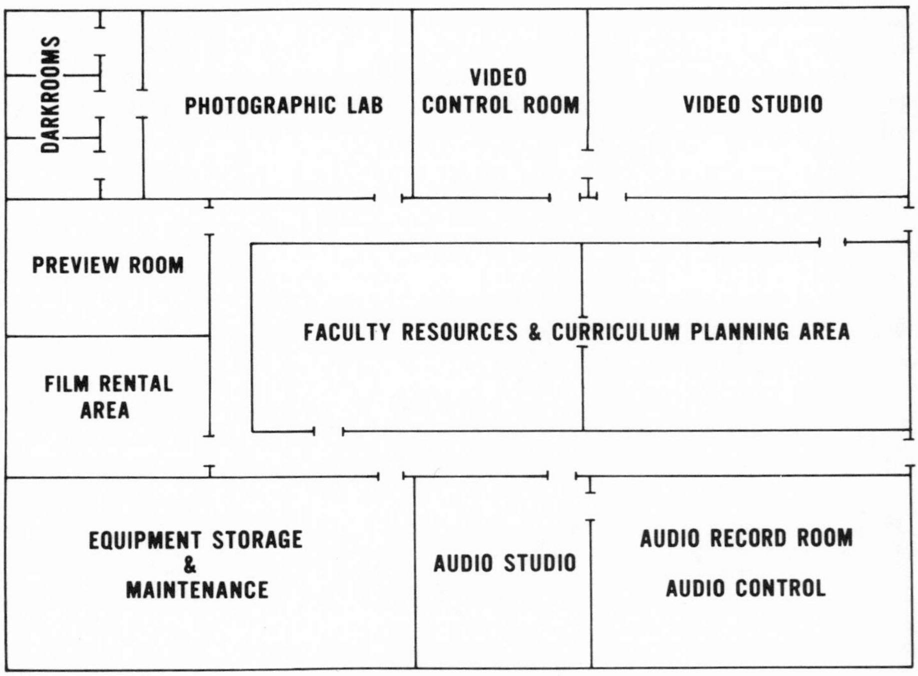

In 1975, the American Library Association and the Association for Educational Communications and Technology issued the publication entitled *Media Programs: District and School*, which delineated guidelines and recommendations for media programs and resources. Although the guidelines are intended for school programs facilities, the application of these recommendations can be helpful when planning AV facilities in other settings. For example, the recommendation that the equipment storage and distribution area be located near an elevator and corridor is an excellent consideration for all AV facilities. Such an arrangement affords the best opportunity for easy access and mobility of equipment. Again, final determinations will depend upon the space available, the purposes, and the needs of the center. Recommendations for school program facilities based on the needs of a school with 1,000 (or fewer) students were as follows:[6]

Areas	Relationships and Special Considerations	Suggested Space Allocations
Circulation (for display, exhibits, copying equipment, card catalogs, periodical indexes, charging)	RELATIONSHIPS. Near main entrance; near reserve collection; near work area; near equipment storage area	800 sq. ft. 200 sq. ft. additional for circulation area in each satellite center
Reading, Browsing, Listening, Viewing	RELATIONSHIPS. Reference area near card catalog, periodical indexes. Magazines near periodical indexes, microform readers.	15 to 30 percent of enrollment at 40 sq. ft. per student A minimum of 9 sq. ft. of floor space is required per single carrel. Some carrels may take 15 sq. ft.
Open Access Materials Housing (Usually integrated into reading, browsing, listening, and viewing areas)	RELATIONSHIPS. Various media formats may be housed on separate shelves or in storage cabinets or interfiled on open shelves according to subject. Reserve area may be open or closed and should be near the circulation area.	Shelving and/or cabinets to accommodate a minimum of 40 items per student, exclusive of textbooks
Small Group Listening and Viewing	RELATIONSHIPS. Small group listening and viewing may be accommodated in open areas of the media center via use of headsets, rear-screen projection, etc. Additional small group listening and viewing areas may be necessary.	Minimum of 150 sq. ft. per area
Conference Areas	RELATIONSHIPS. Locate in quiet area of media center. Consider housing here special collections of materials for which continuous access is not required.	Minimum of 3 conference rooms of 150 sq. ft. each
Group Projects and Instruction	RELATIONSHIPS. Adjacent to reference and open materials housing, and to catalogs and indexes, if possible.	900-1200 sq. ft.
Administration	RELATIONSHIPS. Office for head of the media program should be near the professional collection and easily accessible from rest of school.	Desk space for media staff as necessary; 150 sq. ft. per media professional

(Continues on page 58)

Area (cont'd)	Relationships and Special Considerations (cont'd)	Suggested Space Allocations (cont'd)
Work Space	RELATIONSHIPS. Locate near production and distribution facilities and equipment storage, when possible. Needs access to corridor and to elevator or loading dock.	300-400 sq. ft.
Equipment Storage and Distribution	RELATIONSHIPS. Locate near corridor and freight elevator. Consider location in relation to production area and staff work space.	Minimum of 400 sq. ft.
Maintenance and Repair	RELATIONSHIPS. Near freight elevator, loading dock; adjacent to equipment storage and distribution area.	120-200 sq. ft.
Media Production Laboratory	RELATIONSHIPS. Consider locating adjacent to equipment storage and distribution area.	Minimum of 800 sq. ft. Additional space in schools in which students produce materials
Darkroom	RELATIONSHIPS. The darkroom area, if included in the media center, should be adjacent to the media production laboratory. (A darkroom may be provided elsewhere in the school.)	150-200 sq. ft.
Professional Collection for Faculty	RELATIONSHIPS. Consider in relation to location of teacher's lounge, media production laboratory, department offices, main media center.	Minimum of 600 sq. ft.
Stacks	RELATIONSHIPS. Locate near reserve area, if appropriate. Consider location in relation to periodical storage.	Minimum of 400 sq. ft.
Magazine and Newspaper Storage	RELATIONSHIPS. Locate near periodical indexes, current periodical shelving, and microform readers.	Minimum of 400 sq. ft.
Computerized Learning Laboratory	RELATIONSHIIPS. Adjacent to group project and instruction area.	Depends on nature of computer
Radio Studio	RELATIONSHIPS. May be located near television production area (if any).	500 sq. ft. (20x25 feet) with additional space for control

Areas (cont'd)	Relationships and Special Considerations (cont'd)	Suggested Space Allocations (cont'd)
Television Studio	RELATIONSHIPS. Should be convenient to media production area.	1600 sq. ft. studio (40x40 feet); additional control space; 15-foot ceilings, 14x 12-foot doors
Television Storage	RELATIONSHIPS. Adjacent to television studio.	Minimum of 800 sq. ft.
Television Office	RELATIONSHIPS. Adjacent to television studio and storage areas.	150 sq. ft.

The spatial relationships of individual areas in the audiovisual facility of the learning center can be seen in the diagram in Figure 2-8.

Figure 2-8. Spatial Relationships of Learning Center Services

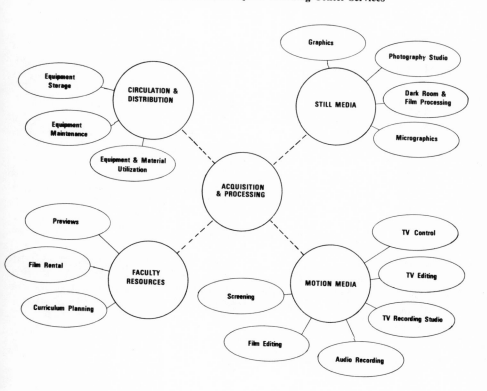

TYPES OF FURNITURE

The type of furniture necessary for a learning center will depend upon the needs of the organization and its program activities. It is important to point out that the selection and acquisition of the appropriate furniture for the facility is extremely important. This alone is the major factor in determining the type of environment that will foster effective learning and utilization. Furniture generally found in a learning center includes the following:

Carrels. Two common types are dry and wet. The dry carrel provides study space for an individual user. The wet carrel provides an electrical capability for audiovisual equipment. These may be arranged in a variety of patterns.

Tables. Study tables both for individuals and for group activities; reference and periodical index tables; production and work areas tablews for projects; lounge area tables and typing tables.

Staff furniture. Staff work stations include desks and chairs, stools of counter height, and typing tables.

General furniture. Storage cabinets for housing materials, supplies, and equipment; shelving for housing books, materials, and equipment; circulation desk; card catalog; display shelves and bulletin boards; filing cabinets for vertical files; atlas and dictionary stands.

NOTES

[1]Association for Educational Communications and Technology, *Guidelines for Certification of Media Specialists.* (Washington, DC: AECT, 1977), pp. 28-29.

[2]American Library Association, *Library Education and Personnel Utilization* (Chicago: ALA, 1970), p. 4.

[3]American Library Association, *Library Education,* p. 2.

[4]American Library Association, "Criteria for Programs to Prepare Library/Media Technical Assistants," *American Libraries* 2 (November 1971), p. 1060.

[5]American Library Association, *Library Education,* p. 4.

[6]Reprinted by permission of the American Library Association and the Association for Educational Communications and Technology from *Media Programs: District and School,* copyright © 1975 by the American Library Association and the Association for Educational Communications and Technology.

BASIC SOURCES

American Library Association. Library Education Division. "Criteria for Programs to Prepare Library/Media Technical Assistants." *American Libraries* 2: 1059-1063 (November 1971).

American Library Association. Library Education Division. "Library Education and Manpower: ALA Policy Proposal." *American Libraries* 1: 341-44 (April 1970).

Association for Educational Communications and Technology. *College Learning Resources Programs: A Book of Readings.* Washington, DC: AECT, 1977.

Association for Educational Communications and Technology. *Training Programs for Educational Media Technicians.* Washington, DC: AECT, 1972.

Association for Educational Communications and Technology. Certification Committee. *Guidelines for Certification of Media Specialists.* Washington, DC: AECT, 1977.

Evans, Charles. *Paraprofessional Library Employees.* Cleveland, OH: Council on Library/Media Technical-Assistants, 1977.

Garoogian, Rhoda. *The Library Technical Assistant.* New York: Pratt Institute, 1975.

Gill, Suzanne, comp. *A Bibliography on Library/Media Technicians.* Cleveland, OH: Council on Library/Media Technical-Assistants, 1978.

Media Programs: District and School. Prepared jointly by American Association of School Librarians, American Library Association, and Association for Educational Communications and Technology. Chicago: ALA; Washington, DC: AECT, 1975.

Merrill, Irving R., and Harold A. Drob. *Criteria for Planning the College and University Learning Resources Center.* Washington, DC: AECT, 1977.

Moffitt, Gerald J. *New Careers Project Lattice.* Toledo, OH: University of Toledo, 1970.

Peterson, Gary. *The Learning Center: A Sphere for Nontraditional Approaches to Education.* Hamden, CT: Linnet Books, 1975.

Roney, Raymond G., and Audrey V. Jones, ed. *1981 Directory of Institutions Offering Programs for the Training of Library/Media Technical Assistants,* 5th ed. Des Peres, MO: Council on Library/Media Technical Assistants, 1981.

Finnish Bookshop, ca 1880

3 – ACQUISITION AND ORGANIZATION
OF AUDIOVISUAL MATERIALS

The acquisition and organization of media are vital activities within the audiovisual center. In order to achieve the objectives of the center, it will be necessary to select, procure, and organize the resources deemed appropriate for that facility. These activities make it possible for the center to acquire the desired resources selectively, to provide the means of locating and retrieving these resources, and to prepare the resources physically for use by the patrons.

Although the selection of materials for the audiovisual center is generally the responsibility of the specialist, the technicians very often assist with the many facets of accomplishing this task. Assignments may include copying selected lists of materials found in books and periodicals or typing those compiled by the specialist in order to obtain user participation in the selection process.

EVALUATION CRITERIA

The procedures for selecting materials differ from one AV center to another. Generally, the center will have a well-thought-out selection policy reflecting the philosophy, goals, objectives, and needs of the parent organization. One aspect of this selection policy will comprise the criteria for the evaluation of media. Usually, the criteria will include the following:

1) **Accuracy.** Is it accurate and current?
2) **Scope.** Is it appropriate for the audience intended? Is the subject matter appropriate?
3) **Style.** Is the presentation well-organized and balanced; imaginative and original?
4) **Interest.** Will it capture and retain the interest of the intended audience?
5) **Technical values.** Is the production of such quality that it presents good color, sound, editing, focus, synchronization, and composition?
6) **Utilization values.** Is it durable and convenient to use and handle? What are its utilization possibilities? Will it add to or complement existing items in the collection?

Direct reviewing, previewing, and auditioning continue to be the best methods of selection for an audiovisual collection. While this may be the ideal way of selecting such materials, it is often difficult to do so due to lack of time available for previewing and auditioning and to the many difficult problems encountered with the procurement and return of such materials. Thus, the general practice of relying on selection aids, such as the producer's recommendations and advertisements (including catalogs and recommendations by sales representatives which are generally noted and filed separately for later consultation), and on evaluative reviewing sources is often the best approach.

Using the items included in the selection criteria for materials to be acquired by the center, an evaluation form should be designed and completed indicating the acceptability of the material and designating whether it is recommended for purchase. The evaluation form may vary from a simple 3x5-inch card format to a more comprehensive detailed form. An example of a rather simple form is shown in Figure 3-1 on page 64.

SELECTION AIDS

The use of reviews as a selection tool is a valuable technique for selecting materials. This method makes possible a broader coverage and critical evaluations of materials by experts with subject and technical specialization.

Unlike print media, only a small percentage of the nonprint media produced each year is reviewed. Although audiovisual reviews vary in quality and scope—some give in-depth evaluations and others give brief descriptive notes—the following sources will be helpful in assisting with the routines involved in the selection of these materials:

Periodicals

Audiovisual Instruction. Department of Audiovisual Instruction.
National Education Association
1201 16th Street, NW
Washington, DC 20036
Published monthly. Includes descriptive reviews of new audiovisual materials.

Figure 3-1. Sample AV Materials Evaluation Form

UNIVERSITY OF THE DISTRICT OF COLUMBIA
LIBRARY AND MEDIA SERVICES
EVALUATION OF INSTRUCTIONAL MATERIAL

Date ____/____/____

Title _____ Series _____

Type of material _____

          Frames _____   B&W   Silent
          Time   _____   Color   Sound Sync.

Publisher _____ Address _____
. . . .

1. Does the material present information related to your instructional goals?
    Excellent   Good   Acceptable   Poor

2. Is the material designed to communicate effectively to students of the age
    and grade level for which its subject matter is appropriated?
    Excellent   Good   Acceptable   Poor

3. Does the material complement information presented about the same subject
    in currently available textbooks and other media?
    Excellent   Good   Acceptable   Poor

4. For what curricular or University use would this film be suitable? _____
    _____
    _____

5. Technical Quality (sound, color, film)
    Excellent   Good   Acceptable   Poor

6. Anticipated use of material
        Independent Study _____   Small Group _____   Lecture _____
        Seminar   _____   Large Group _____   Other   _____

7. Will this material become outdated in near future?   Yes _____ No _____
    If yes, Explain. _____
    _____

8. Would you recommend purchase?   Yes _____ No ___   Rental? Yes _____ No __

9. Justification. _____
    _____

10. Name of Evaluator _____ Position _____

11. Department _____

Booklist. American Library Association
   50 E. Huron Street
   Chicago, IL 60611
   Published semimonthly. Provides reviews of filmstrips, disc and cassette recordings, 16mm and 8mm films, videocassettes, transparencies, and multimedia kits.

Film News. Film News Company
   250 W. 57th Street
   New York, NY 10019
   Published bimonthly. Includes descriptive and annotated evaluations of 16mm films, filmstrips, and phonodiscs.

Hi Fidelity/Musical America. Billboard Publications
   Great Barrington, MA 01230
   Published monthly. Provides reviews and lists of musical recordings, audiotapes, and discs.

The Instructor. Instructor Publications Inc.
   Instructor Park
   Dansville, NY 14437
   Published monthly. Provides reviews of filmstrips, films, charts, kits, and other curriculum materials.

Media & Methods. North American Publishing Co.
   134 North American Building
   401 N. Broad Street
   Philadelphia, PA 19108
   Published monthly from September through May/June. Includes some reviews of filmstrips, films, audiocassettes, and various other media.

Media Review Digest. Pierian Press
   5000 Washtenaw Avenue
   Ann Arbor, MI 48104
   Published annually with semiannual supplements. An index to the digest of nonbook media reviews, evaluations and descriptions appearing in a great variety of periodicals and reviewing services.

Previews. R. R. Bowker Co.
   1180 Avenue of the Americas
   New York, NY 10036
   Published monthly from September through May. Includes reviews of art prints, nonmusical recordings, 16mm films, filmstrips, and multimedia kits.

School Library Journal. R. R. Bowker Co.
   1180 Avenue of the Americas
   New York, NY 10036
   Published monthly. Provides lists and reviews of recording films, filmstrips, and other media useful in schools and public libraries.

Source Guides

The NICEM Index series, compiled from information stored in the National Information Center for Educational Media at the University of Southern California (University Park, Los Angeles, CA 90007), gives access to over 300,000 titles of educational media. Includes subject guides, individual and series titles, and descriptive notes. Also provides a directory of producers and distributors.

NICEM Index to Educational Audio Tapes. 1980.
NICEM Index to Educational Overhead Transparencies. 1980.
NICEM Index to Educational Records. 1980.
NICEM Index to Educational Slides. 1980.
NICEM Index to Educational Videotapes. 1980.
NICEM Index to 8mm Educational Motion Cartridges. 1980.
NICEM Index to 16mm Educational Films. 1980.
NICEM Index to 35mm Educational Filmstrips. 1980.

Film Evaluation Services

Film evaluation services are also available on a subscription basis. Some of the more popular ones are:

Educational Film Locator. R. R. Bowker Co.
1180 Avenue of the Americas
New York, NY 10036
Union list of titles held by member libraries of the Consortium of University Film Centers, and a compilation and standardization of their 50 separate catalogs, representing about 200,000 film holdings with their geographic locations.

EFLA Evaluations. Educational Film Library Association, Inc.
16 W. 60th Street
New York, NY 10023
Published monthly. Available to members only. Provides evaluations of films in various instructional subjects for all age levels. Alphabetically listed by title with critical comments and general ratings.

Landers Film Reviews. Landers Associates
P.O. Box 69760
Los Angeles, CA 90069
Published bimonthly, September through May. Comprehensive evaluations of current 16mm films. Arranged in Alphabetical order with subject and title indexes.

National Union Catalog. 1956- . Washington, DC: Library of Congress
An important source for cataloging and bibliographic information for acquisition, reference, and research. Issued in three quarterly issues, with annual and quinquennial cumulations. The *Films and Other Materials for Projection* (1973-) and *Music, Books on Music and Sound Recordings* (1973-) are the related sources for nonprint materials. Recently, *Audiovisual*

Materials (1980-) was issued for motion pictures, filmstrips, sets of transparencies, slide sets, videorecordings, and kits currently cataloged by the Library of Congress.

Nonprint Media Review Indexes

Indexes to reviews of nonprint media are also helpful selection aids. Two examples of such indexes are:

Media Review Digest. Pierian Press
5000 Washtenaw Avenue
Ann Arbor, MI 48104
Published annually. Reviews, evaluations, and descriptions of nonbook media found in periodicals and reviewing services.

Notes. Music Library Association, Research Library of the Performing Arts
111 Amsterdam Avenue
New York, NY 10023
Published quarterly. Reviews of recordings in various periodicals are indexed with notations indicating favorable or unfavorable reviews.

Producers' Catalogs

A collection of producers' catalogs should be actively developed and maintained to aid in selection. These catalogs can be acquired by sending requests to the major companies. The mailing information can be located by consulting the lists of distributors and producers in many of the selection aids previously cited.

ACQUISITION OF MATERIALS

Many facets of the routines necessary for ordering and buying materials are often assigned and completed by the library/media technical assistant. For the media technician, these routines usually include the following:

1) Receiving and typing order requests
2) Searching the catalog and outstanding-order file
3) Verifying information
4) Typing and mailing orders
5) Filing in the outstanding order files
6) Receiving and checking in materials
7) Acquiring approval for payment and maintaining records of payment
8) Forwarding materials for cataloging

Order Requests

Requests for orders are submitted by the users or selectors in various ways. Some centers accept requests in any fashion that is legible, such as publication announcements, brochures, producers and distributors catalogs, and handwritten or typewritten notes submitted by the selector. Other centers require an order-request form to be completed for each item requested for purchase. The format may vary, but generally the same request form for books is used for all media (*see* Figure 3-2).

Figure 3-2. Sample Request Form

Class Number	Author's (surname first)
Number of Copies	Title
Date Ordered	Volumes
Of	Edition Year Publishers List Price
Received	Requested by
	Notify Address
Date of Bill	Reviewed in
Cost per copy	Approved by

Usually the form will be a 3x5-inch multi-part order card that elicits information about selections the selector might not otherwise include. In addition, the various parts of the form may be arranged in a variety of files as deemed necessary.

Searching the Catalog and the Outstanding-Order File

After the receipt of the order request, usually the technician will check the center's catalog to determine whether or not the item is owned by the center. In some instances, it may be necessary for the searcher to do a comprehensive search to discover whether the title is part of a series that is cataloged under the names of the authors or producers in the series. The technician should also check the outstanding-order file where the center has on file all materials that are on order and in process and therefore not in the center's catalog. Although some centers file copies of outstanding orders in the catalog, generally, these cards or slips are maintained in a separate file.

The process of checking the information submitted by a selector for accuracy and completeness is known as verification. If an item requested for purchase is not found in the catalog or in the outstanding-order file, a search for verification of information is undertaken. This is done by searching in one or more of the indexes and bibliographic lists previously mentioned. Once found, the information is checked and any additional information not included on the order request form should be added. Figure 3-3 shows an example of a modified book form for audiovisual materials.

Figure 3-3. Materials Order Form

CLASS NO.	STATUS		PURCHASE ORD. #	L.C. CARD NO.	
	OP	OS			
ACCT. #	TITLE				
NO. COPIES	TYPE OF MATERIAL				
LIST PRICE	PRODUCER		DATE		
COST PRICE	ACCESSION NO.	FUND CHARGED		SPECIAL INSTRUCTIONS	
DATE ORDERED	RECOMMENDED		LIBRARY NAME		
CANCEL					

Once the information has been verified and the order form completed, the next step is the placing of the order. In some centers, a requisition form (*see* Figure 3-4, page 70) is prepared and directly sent to the vendor. Other centers have purchasing offices whose responsibility is to place the orders directly with the vendors. The requisition form generally includes a purchase order number, unit cost, discount cost, and invoicing and shipping instructions.

An invaluable file to organize and maintain is that of the producers catalogs. As the catalogs are received, they should be filed for ready reference and quick retrieval in ordering materials. An accompanying producers card file is also maintained in many centers. Information essential for ordering materials from a company is recorded and alphabetized by company on 3x5-inch cards. This data will include: the company's name and address, the types of material supplied, and pertinent information about the local representative, etc. Both files should be continually updated with old catalogs replaced as the new ones are received. An excellent source of distributor information is *Audiovisual Marketplace: A Multi-Media Guide* (R. R. Bowker Co., 1180 Avenue of the Americas, New York, NY 10036). Published annually, it provides producer and distributor information classified by media and subject. If Library of Congress cards are to be ordered for many of the materials, the National Union Catalog will assist in providing LC card numbers.

Once the order has been placed, the multi-part request/order form is then separated and the various parts are filed in one or all of the following: the purchase/order file, the outstanding-order file; the purchasing office files (with requisition form). One copy may also be retained by the requestor.

Figure 3-4. Materials Requisition Form

UDC-300   REQUISITION

UNIVERSITY OF THE DISTRICT OF COLUMBIA

AGENCY DOCUMENT NUMBER

NO. **3499**

Entry [X] REQE
Modify [ ] REQM
Cancel [ ] REQX

LINE NO.	AGY	RESP CTR PROJ- PHASE	MRU SUB- PROJ	OBJECT/ SUBOBJECT	DIST	AGCY	JOB NO.	AC- TIV ITY	+ / -	TOTAL AMOUNT
						REPT CATG				
0001	GF	2013		708			U520			$157.00

DATE APPROVED
Mo Dy Yr

DATE NEEDED
Mo Dy Yr

DELIVER TO: _____ NAME _____ BLDG/ROOM _____

PREPARED BY:

DATE 11-20-79

AUTHORIZED BY:

NAME (Print) _____ SIGNATURE _____ PHONE _____ DATE _____

CERTIFICATE OF ACCEPTANCE INTO FMS.   I certify that this document was accepted into FMS.

CHECK IF DOCUMENT IS CONTINUED PAGE ___ OF ___

Signature _____

Date Accepted _____

ISSUE FROM INVENTORY

STOCK NUMBER	DESCRIPTION	QUAN	UNIT	UNIT PRICE	TOTAL PRICE	REMARKS SUGGESTED VENDORS
	16MM Sound Color Film, 11 minutes					Aims Instructional Media Services, Inc. 626 Justin Avenue Glendale, Calif. 92201
	BLACK DIMENSIONS IN AMERICAN ART	1	ea.	155.00	$ 155.00	
	Shipping & Handling				2.00	

Receiving

After the material is received, the item is matched to the purchase order to determine whether the items received correspond to items ordered, whether the quantity and price are correct, and whether the materials are in good condition. If damage is discovered, the materials should be returned immediately. A notation should be made on the purchase order of the date of arrival. Once the materials received are determined to be in agreement with the purchase order, the invoice is approved and forwarded to the purchase office for payment. A record of payment and a record of unfilled orders are maintained. Finally, when the ordering and receiving procedures are completed, the material and a copy of the purchase order slip are forwarded for cataloging.

ORGANIZATION AND ARRANGEMENT

Upon receipt of the materials, it is necessary that they be organized and processed so that they can be easily identified and retrieved by the users. In order to identify the materials and to inform the users of their availability, the center will have to establish a classification system by which they are to be organized and maintained. While the practice of separating print and nonprint media into separate collections is utilized by many centers, the integration of both resources permits access to all media at the same time. Further, the decision to include both print and nonprint entries in one catalog (or to maintain a separate catalog for print and nonprint media) is also of major importance.

Although the techniques used in the organization of audiovisual materials vary to a greater extent with nonprint than with print materials, there appear to be three classification systems in general use. The first is the assignment of an accession number for each item received and added to the collection. This practice is more widely used in small AV centers. The second is the Dewey Decimal Classification system, which is used more often by centers in elementary and secondary schools. The third is the Library of Congress Classification system, which is designed for larger, more extensive collections. Often, a symbol indicating the type of material is written above the call number for location and storage purposes.

Symbol Designation and Codes

The inclusion of a symbol with the accession or call number is often practiced in cataloging audiovisual materials to distinguish the types of materials added and their locations within the collection. The practice of spelling out the type of material for ease of recognition is followed by many centers; however, the use of abbreviated codes and symbols is widely practiced also. These symbol designations are placed above the classification numbers and thus become part of the call numbers (*see* Figure 3-5, page 72).

Figure. 3-5. Symbol Designations as Part of Call Numbers

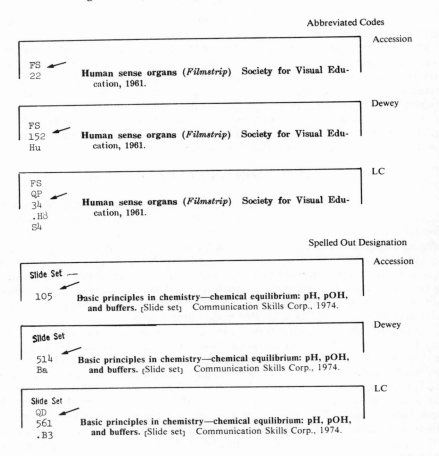

In AECT's *Standards for Cataloging Nonprint Materials*, the following types of media and their symbols are recommended:

Medium Designator	Specific Designator	Code
Audiorecording..		AA
	Cartridge	AR
	Cassette	AC
	Disc	AD
	Reel	AT
	Roll	AO
Archival/Experimental..........	Cylinder	AY
	Page	AS
	Wire	AW

Medium Designator (cont'd)	Specific Designator (cont'd)	Code (cont'd)
Chart..		CA
	Chart	CH
	Flannel board set	CL
	Flip chart	CF
	Graph	CG
	Magnetic board set	CM
	Relief chart	CR
	Wall chart	CW
Diorama..		OA
	Diorama	OD
Filmstrip...		FA
	Filmslip	FL
	Filmstrip	FS
Flash card.......................................		HA
	Card	HC
Game...		GA
	Game	GM
	Puzzle	GP
	Simulation	GS
Globe..		QA
	Globe	QG
	Relief globe	QR
Kit..		KA
	Exhibit	KE
	Kit	KT
	Laboratory kit	KL
	Programed instruction kit	KP
Machine-readable data file....................		DA
Species of file	Date file	DF
	Program file	DE
Storage medium	Disc	DD
	Punched card	DB
	Punched paper tape	DP
	Tape	DT
Map..		LA
	Map	LM
	Relief map	LR
	Wall map	LW
Microform.......................................		NA
	Aperture card	NC
	Card	ND
	Cartridge	NE
	Cassette	NF
	Fiche	NH
	Reel	NR
	Ultrafiche	NU
Model..		EA
	Mock-up	EM
	Model	EE

Medium Designator (cont'd)	Specific Designator (cont'd)	Code (cont'd)
Motion picture		MA
	Cartridge	MR
	Cassette	MC
	Loop	ML
	Reel	MP
Picture		PA
	Art original	PO
	Art print	PR
	Hologram	PH
	Photograph	PP
	Picture	PI
	Post card	PC
	Poster	PT
	Stereograph	PG
	Study print	PS
Realia		RA
	Name of object	RO
	Specimen	RS
Slide		SA
	Audioslide	SO
	Microscope slide	SM
	Slide	SL
	Stereoscope slide	SS
Transparency		TA
	Transparency	TR
Videorecording		VA
	Cartridge	VR
	Cassette	VC
	Disc	VD
	Reel	VT

Each medium may be coded either by the general form designator—e.g., HA for Flash Card (which includes card)—or by the specific physical form designator (which cites a separate coding for card: HC) for a more specific designation of the type of material.

The practice of identifying different forms of material through the use of color codes is currently not recommended. Such a system is more costly in time and materials because of the variety of color cards and labels needed to maintain this system.

Classification Schemes

The use of classification schemes provides a means of bringing materials together on the same subject for greater accessibility and utilization by users. Classification, then, is a process of arranging items, objects, or ideas into groups according to their intellectual content.

Accession Number Systems

The use of a number in order to identify each item added to the collection in a chronological order is called the accession number system. Generally, the numbers begin with 1 (for the first item received) and in some cases climb into the thousands. Materials of the same type may be grouped together under a specific symbol and a number assigned to each item as it is received within the group—e.g., Motion Picture—MP-1 and Art Print—PA-1. Such groupings permit items of the same type to be placed sequentially on the shelves.

This system requires very little decision making and is thereby a simple method of classifying materials. Figure 3-6 shows examples of various types of materials classified with the accession number system:

Figure 3-6. Cards for Materials Classified with Accession Number System

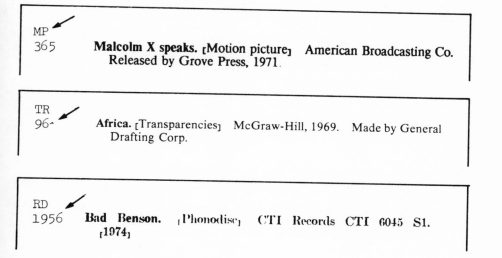

MP
365   **Malcolm X speaks.** [Motion picture]   American Broadcasting Co. Released by Grove Press, 1971.

TR
96   **Africa.** [Transparencies]   McGraw-Hill, 1969.   Made by General Drafting Corp.

RD
1956   **Bad Benson.**   [Phonodisc]   CTI Records CTI 6045 S1. [1974]

Dewey Decimal Classification System

The classification scheme used in most libraries and audiovisual centers in elementary and secondary schools, as well as in small centers in public and private organizations and institutions, is the Dewey Decimal Classification system. Devised by Melvil Dewey and first published in 1876, it is based on 10 main classes, of which 9 cover areas of knowledge that can then be divided into 9 smaller parts that can also be subdivided, and then further subdivided ad infinitum.

Both the Dewey Decimal and the Library of Congress Classification systems range from the general to the specific. In summary form, the main classes and subdivisions of the Dewey decimal system are:*

000 Generalities
010 Bibliographies and catalogs
020 Library science
030 General encyclopedic works
040 Essays, addresses, lectures
050 General periodicals
060 General organizations
070 Newspapers and journalism
080 General collections
090 Manuscripts and book rarities

100 Philosophy and related
110 Ontology and methodology
120 Knowledge, cause, purpose, man
130 Pseudo- and parapsychology
140 Specific philosophic viewpoints
150 Psychology
160 Logic
170 Ethics (Moral philosophy)
180 Ancient, medieval, Oriental
     philosophy
190 Modern Western philosophy

200 Religion
210 Natural religion
220 Bible
230 Christian doctrinal theology
240 Christian moral and devotional
     theology
250 Christian pastoral, parochial, etc.
260 Christian social and ecclesiastical
     theology
270 History and geography of Christian
     church
280 Christian denominations and sects
290 Other religions and comparative

300 The social sciences
310 Statistical method and statistics
320 Political science
330 Economics
340 Law
350 Public administration
360 Welfare and association
370 Education
380 Commerce
390 Customs and folklore

400 Language
410 Linguistics and nonverbal language
420 English and Anglo-Saxon
430 Germanic languages
440 French, Provencal, Catalan
450 Italian, Romanian, etc.
460 Spanish and Portuguese
470 Italic languages
480 Classical and Greek
490 Other languages

500 Pure sciences
510 Mathematics
520 Astronomy and allied sciences
530 Physics
540 Chemistry and allied sciences
550 Earth sciences
560 Paleontology
570 Anthropological and biological
     sciences
580 Botanical sciences
590 Zoological sciences

600 Technology (Applied science)
610 Medical sciences
620 Engineering and allied operations
630 Agriculture and agricultural industries
640 Domestic arts and sciences
650 Business and related enterprises
660 Chemical technology, etc.
670 Manufactures processible
680 Assembled and final products
690 Buildings

700 The arts
710 Civic and landscape art
720 Architecture
730 Sculpture and the plastic arts
740 Drawing and decorative arts
750 Painting and paintings
760 Graphic arts
770 Photography and photographs
780 Music
790 Recreation (Recreational arts)

---

*Reprinted from *Dewey Decimal Classification*, 19th edition, © 1979 by permission of Forest Press Division, Lake Placid Education Foundation, owner of copyright.

800 Literature and rhetoric	900 General geography and history, etc.
810 American literature in English	910 General geography
820 English and Anglo-Saxon literature	920 General biography, genealogy, etc.
830 Germanic languages literature	930 General history of ancient world
840 French, Provencal, Catalan literature	940 General history of modern Europe
850 Italian, Romanian, etc., literature	950 General history of modern Asia
860 Spanish and Portuguese literature	960 General history of modern Africa
870 Italic languages literature	970 General history of North America
880 Classical and Greek literature	980 General history of South America
890 Literatures of other languages	990 General history of rest of world

Figure 3-7 shows examples of various types of materials classified with the Dewey classification system:

Figure 3-7. Cards for Materials Classified with Dewey Decimal System

---

Filmstrip

649.3
F295   **Feeding your young children** (*Filmstrip*)   National Dairy Council, 1968.

MP
371.33   **To help them learn.** ₍Motion picture₎ / Association for Educational
T627   Communications and Technology, Association of Media Producers. — Washington : Association of Media Producers, 1977.

Slide Set

544
B311   **Basic NMR spectroscopy: NMR application in industrial, clinical, and special laboratories.** ₍Slide set₎   Communication Skills Corp., 1975.

TR
500.2   **Earth science transparencies, set 1.** ₍Transparencies₎   McGraw-
Ea 12   Hill, 1970.   Made by Kieffer-Nolde Offset Printing.

Library of Congress Classification System

The Library of Congress Classification was devised for the national library bearing its name. The system is used by many large public and college and university libraries. Some small libraries with extensive subject collections and some audiovisual or learning centers have also adopted the scheme.

This system places all knowledge into 21 major classes and combines letters of the alphabet with arabic numbers to form the classification number. Presently, the letters I, O, W, X, and Y are not used and are reserved for further expansion. The following is a general outline of the letter classes:

A   General Works

B   Philosophy. Psychology. Religion

C   Auxiliary Sciences of History (General—Civilization, Archeology, Diplomatics, Genealogy, etc.)

D   History—General and Old World

E   American History & General U.S. History

F   American History (local) & Latin American

G   Geography. Anthropology. Recreation

H   Social Sciences

J   Political Science

K   Law

L   Education

M   Music and Books on Magic

N   Fine Arts

P   Language & Literature

Q   Science

R   Medicine

S   Agriculture

T   Technology

U   Military Science

V   Naval Science

Z   Bibliography. Library science

Figure 3-8 shows examples of various types of materials classified with the Library of Congress Classification system:

Figure 3-8. Cards for Material Classified with Library of Congress System

---

Motion Picture

PN
1997   **The Perfect race.** ₍Motion picture₎   Directions Unlimited Film
.P4   Corp.   Released by Pyramid Films, 1972.

TR
RC
667   **The Patient and circulatory disorders.** ₍Transparencies₎   J. B. Lip-
.P3   pincott Co., 1969.

Kit
LB
2395   **How to survive in school—note-taking and outlining skills.** ₍Slide₎.
.H6   — White Plains, N.Y. : Center for Humanities, 1977, made
            1976.

Call Number

The classification symbol, accompanied by the title or author number, and the title work mark is known as the *call number*, which is placed on a catalog card to show the location of an item. It generally appears in the upper left-hand corner of the card as shown in Figure 3-9. The call number is also placed in some predetermined location on the item itself.

Figure 3-9. Call Number Location on a Catalog Card

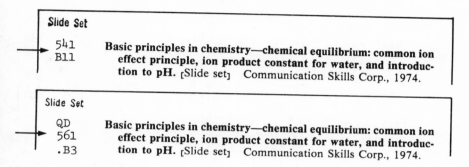

In addition to the classification number, which is the first part of the call number, a *title* or *author number* is also assigned. Not all centers or libraries use author numbers; some use only the first letter of the author's name. The title or author number is derived by combining the first letter of the title or author's last name with one- , two- , or three-figure numbers. This is generally known as the *Cutter number* and based on a table of numbers and letters developed by C. A. Cutter. The *Cutter Two-Figure Table* is a four-page table primarily designed for small collections. The *Cutter Three-Figure Table* is a 26-page table designed primarily for use in medium-sized collections. The *Cutter-Sanborn Three-Figure Table* is an 18-page table used in larger collections (*see* Figure 3-10, page 80). The combination of the initial letter of the title or author's name followed by a number will place the title or author in an alphabetical position in relation to other titles or authors with names beginning with the same initial letters.

Although less frequently used, an additional letter can be added to further arrange the items alphabetically by the title, particularly within a series, as well as by author. This is known as the *work mark*, and it is derived by using the first letter of the first word in the title (exclusive of articles) when using the author number, and of a title within a series when using the title number. This is utilized more with extensive collections in which one author has more than one title on the same subject and which include collections of materials in series (*see* Figure 3-11, page 81).

Figure 3-10. Cutter-Sanborn Sample

CUTTER-SANBORN Three-Figure Author Table (SWANSON-SWIFT REVISION)

Ba	111	Bamp	211	Basi	311	Bede	411	Beri	511
Bab	112	Ban	212	Basili	312	Bedi	412	Berk	512
Babe	113	Banc	213	Basin	313	Bedr	413	Berkl	513
Babi	114	Band	214	Basir	314	Bee	414	Berl	514
Babr	115	Bane	215	Bask	315	Beer	415	Berlin	515
Bac	116	Bang	216	Basn	316	Beg	416	Berm	516
Bacci	117	Bani	217	Bass	317	Begi	417	Bern	517
Bach	118	Bank	218	Basse	318	Begu	418	Bernar	518
Bache	119	Bann	219	Basset	319	Beh	419	Bernard J	519
Bachell	121	Bao	221	Bassi	321	Behr	421	Bernard M	521
Bachet	122	Bap	222	Basso	322	Bei	422	Bernard T	522
Bachi	123	Bar	223	Bassu	323	Beis	423	Bernardi	523
Bachm	124	Barag	224	Bast	324	Bek	424	Bernat	524
Baci	125	Baran	225	Baste	325	Bel	425	Berne	525
Back	126	Barat	226	Basti	326	Belan	426	Bernet	526
Bacm	127	Barau	227	Basto	327	Belch	427	Bernh	527
Baco	128	Barb	228	Bat	328	Bele	428	Berni	528
Bacon M	129	Barbar	229	Bates	329	Belg	429	Berno	529
Bacr	131	Barbat	231	Bath	331	Beli	431	Berns	531
Bad	132	Barbau	232	Bathu	332	Belk	432	Bero	532
Bade	133	Barbe	233	Bati	333	Bell	433	Berr	533
Baden	134	Barber	234	Bato	334	Bell L	434	Berry	534
Badg	135	Barbet	235	Batt	335	Bell R	435	Bers	535
Badi	136	Barbi	236	Batti	336	Bellan	436	Bert	536
Bado	137	Barbil	237	Bau	337	Bellav	437	Berte	537
Badr	138	Barbo	238	Baud	338	Belle	438	Berth	538
Bae	139	Barbou	239	Baudio	339	Belleg	439	Berthe	539
Baer	141	Barbu	241	Baudo	341	Bellen	441	Berthi	541
Baert	142	Barc	242	Baudr	342	Beller	442	Bertho	542
Baf	143	Barch	243	Baudu	343	Belli	443	Berti	543
Bag	144	Barcl	244	Baue	344	Bellin	444	Bertin	544
Bagi	145	Bard	245	Bauf	345	Bellm	445	Berto	545
Bagl	146	Bardi	246	Baug	346	Bello	446	Bertol	546
Bagn	147	Bardo	247	Baum	347	Bellon	447	Berton	547
Bago	148	Bare	248	Baumg	348	Bellow	448	Bertr	548
Bags	149	Barf	249	Baun	349	Bellu	449	Bertrand F	549
Bah	151	Barg	251	Baur	351	Belm	451	Bertrand N	551
Bai	152	Bari	252	Baut	352	Belo	452	Bertu	552
Bail	153	Barin	253	Bav	353	Belt	453	Berw	553
Baile	154	Bark	254	Bavi	354	Belv	454	Bes	554
Bailey L	155	Barker	255	Bax	355	Bem	455	Besl	555
Bailey S	156	Barki	256	Bay	356	Ben	456	Beso	556
Baill	157	Barl	257	Baye	357	Benc	457	Bess	557
Baillo	158	Barlo	258	Bayl	358	Bend	458	Bessem	558
Bails	159	Barn	259	Bayly	359	Bendo	459	Bessi	559
Baily	161	Barnes	261	Bayn	361	Bene	461	Best	561
Bain	162	Barnh	262	Baz	362	Benede	462	Bet	562
Bair	163	Barnu	263	Bazi	363	Benedi	463	Bethm	563
Bait	164	Baro	264	Bazo	364	Benef	464	Beto	564

Figure 3-11. Card with Call Number Made Up of a Classification Number, a Title Number, and a Work Mark

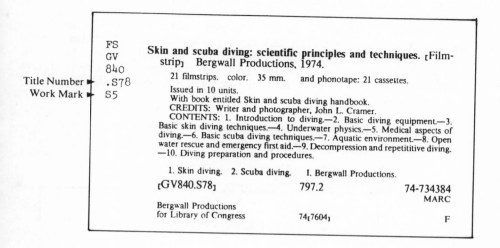

FS GV 840	**Skin and scuba diving: scientific principles and techniques.** [Film-strip] Bergwall Productions, 1974.
Title Number ► .S78 Work Mark ► S5	21 filmstrips. color. 35 mm. and phonotape: 21 cassettes. Issued in 10 units. With book entitled Skin and scuba diving handbook. CREDITS: Writer and photographer, John L. Cramer. CONTENTS: 1. Introduction to diving.—2. Basic diving equipment.—3. Basic skin diving techniques.—4. Underwater physics.—5. Medical aspects of diving.—6. Basic scuba diving techniques.—7. Aquatic environment.—8. Open water rescue and emergency first aid.—9. Decompression and repetititive diving. —10. Diving preparation and procedures.

1. Skin diving. 2. Scuba diving. I. Bergwall Productions.

[GV840.S78]   797.2   74-734384
                                                MARC

Bergwall Productions
for Library of Congress   74[7604]   F

THE CATALOG

Although the card catalog can be found in many centers, other types of catalogs are also utilized by some centers. These include the *book catalog*, in which all entries are listed in a book format; the *microform catalog*, in which entries are either in microfilm or microfiche format; and the *online computer* format, for which a terminal used as a public access catalog is available for the patrons.

The *card catalog* is an index to all materials and their contents housed in a library or audiovisual center. To use it, the patron only needs to know either the author, title, or subject of a specific item. Usually centrally located, it consists of 3x5-inch cards in drawers labelled on the outside with guide cards. Rods inside keep the cards from easily being misplaced and scattered.

Although the *divided catalog* (subject, title, and author filed separately, or author and title in one alphabet and subject in another) is widely used, the *dictionary catalog* is commonly found in many centers and libraries. This arrangement provides for the interfiling of author, title, and subject entries in a single alphabet. The *shelflist catalog* is a different type. It is arranged by call number according to the actual order arrangement of materials on the shelves. There is one entry for each title.

Types of Entries

The name or heading under which an item appears in the catalog is called the *entry heading*. Generally, items are entered under author, title, and subject, but in its *Standards for Cataloging Nonprint Materials* (4th ed.), AECT recommends title entry only for nonbook media.

The two basic individual entry headings found in the catalog are *main entry* headings and *added entry* headings. The main entry heading for a nonbook item may be made under title, series title, or creator. Added entries should be made for those personal or corporate names, original or variant titles, and series titles under which a user might search for the work. Added entry headings for producer/publisher, sponsor, and distributor are not usually made.

For the sake of consistency in form, standard items are placed on the card in a predefined pattern. The sample cards in Figures 3-12 and 3-13 conform to the rules for capitalization, punctuation and the use of numerals as prescribed in the *Anglo-American Cataloguing Rules*, 2nd ed. (Chicago: American Library Association, 1978) (*AACR 2*). The small numbers on the cards indicate the spacing of the elements.

Figure 3-12. Arrangement of Elements on a Catalog Card

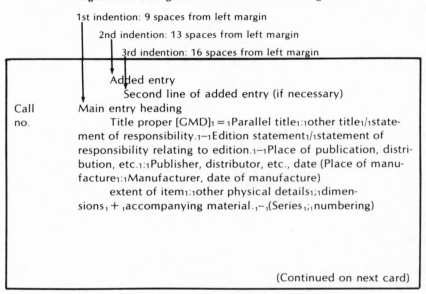

1st indention: 9 spaces from left margin

2nd indention: 13 spaces from left margin

3rd indention: 16 spaces from left margin

Added entry
Second line of added entry (if necessary)

Call Main entry heading
no. Title proper [GMD]₁ = ₁Parallel title₁:₁other title₁/₁statement of responsibility.₁–₁Edition statement₁/₁statement of responsibility relating to edition.₁–₁Place of publication, distribution, etc.₁:₁Publisher, distributor, etc., date (Place of manufacture₁:₁Manufacturer, date of manufacture)
extent of item₁:₁other physical details₁;₁dimensions₁ + ₁accompanying material.₁–₁(Series₁;₁numbering)

(Continued on next card)

Figure 3-13. Sample for Supplementary Typed Card

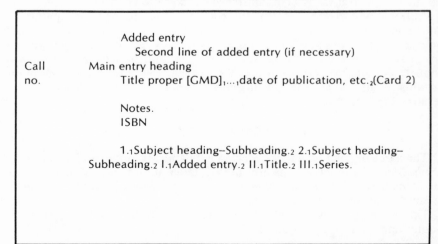

Added entry
Second line of added entry (if necessary)

Call Main entry heading
no. Title proper [GMD]₁...₁date of publication, etc.₂(Card 2)

Notes.
ISBN

1.₁Subject heading–Subheading.₂ 2.₁Subject heading–Subheading.₂ I.₁Added entry.₂ II.₁Title.₂ III.₁Series.

The following are the areas of a bibliographic description as prescribed in *AACR 2*:

TITLE AND STATEMENT OF RESPONSIBILITY AREA

Title proper and alternative title
General material designation (GMD) indicating the type of material being described
Parallel titles (the title of the item in another, applicable language)
Other title information
Statements of responsibility (author, composer, compiler, etc.)

EDITION AREA

Edition statement
Statements of responsibility relating to the edition

MATERIAL (OR TYPE OF PUBLICATION) SPECIFIC DETAILS AREA
(Used only for cartographic materials and serials)

PUBLICATION, DISTRIBUTION, ETC., AREA

Place of publication, distribution, etc.
Name of publisher, distributor, etc.
Date of publication, distribution, etc.
Place of manufacture, name of manufacturer, date of manufacture

PHYSICAL DESCRIPTION AREA

Extent of item
Other physical details
Dimensions
Accompanying material

SERIES AREA

Title proper of series
Statements of responsibility relating to series
ISSN of a series
Numbering within series
Subseries

NOTE AREA

(Notes that qualify or amplify the formal description)

Scope
Language
Title variations
Audience
Other formats available
Contents
Etc.

STANDARD NUMBER AND TERMS OF AVAILABILITY AREA

ISBN, ISSN, or other standard number
Key title of a serial
Price

The completed entry should also include what are called *tracings*, the notations of all the headings under which the item is entered. The completed entries are then arranged in the catalog in alphabetical order according to the first word, excluding initial articles.

Library of Congress Printed Cards

Although not the only source, the Library of Congress in Washington, DC, is the major source for sets of cards for audiovisual materials to be used in the card catalog. Cards may be ordered either by LC card number, by author and title, or by both. The card number for many nonprint media can be obtained by searching the *National Union Catalog* for specific formats of materials, i.e., motion pictures, filmstrips, sound recordings, etc. Once the card for an item is located, the card number can be found in the lower right-hand corner. Using the forms provided by the Library of Congress, or others (*see* Figure 3-3, page 69), the card number and other pertinent information should be provided and the order form mailed to the Library of Congress. If the card number is not available, cards can be obtained by providing the title of the work and other necessary information.

Typed Catalog Cards

When cards are not ordered or are not available (e.g., if no LC cards exist), very often they are prepared in-house. The specialist will do original cataloging and the assistant may be assigned the responsibility of typing the card. If the cards are typed, a standard format is necessary for consistency. The format followed in Figures 3-12 and 3-13, page 82, is as follows:

1. Start the main entry heading on the fourth line from the top of the card.
2. First indention is 9 spaces from the left edge of card.
3. Second indention is 13 spaces.
4. Third indention is 16 spaces.
5. Start call number on sixth line down.

In addition to the main entry card (which is the form of LC printed cards), a card should be included in the catalog at every point the user may be expected to look for the item. These *added entries* are made for coauthors, titles, performers, alternate titles, series, and other designations by which the user may be familiar with the work. (Choice of these *access points* is covered in *AACR 2*, chapter 21). Entries are also made for subjects treated in the item. The form of the subject headings and the cross-references that should be made are prescribed by published subject heading lists. The lists most commonly in use in AV centers and libraries are the *Library of Congress Subject Headings*, now in its 9th edition (Washington, DC: Library of Congress, 1980) and *Sears List of Subject Headings*, 11th ed. (New York: H. W. Wilson, 1977). All of the added entries are typed in the tracings on the main entry card.

An entry for the shelflist should also be completed with information relative to the purchase price, date received, name of vendor, and copy or accession numbers data. Again, the rules for abbreviations, capitalization, punctuation, etc., as described in *AACR 2* should be followed.

Cross-Reference Cards

The cross-reference cards mentioned above direct the user from one heading to another within the card catalog. The *see* cards refer the user from a heading not used to the heading used in the card catalog. The *see also* cards direct the user to additional or related headings after exhausting all references under the heading used (*see* Figure 3-14).

Figure 3-14. Cross-Reference Cards

```
          CIVIL DISOBEDIENCE
          See
          GOVERNMENT, RESISTANCE TO
```

```
          POLITICAL CRIMES AND OFFENSES
            See also
          GOVERNMENT, RESISTANCE TO
```

FILING CATALOG CARDS

Again, although there are other filing schemes, many centers follow the *ALA Rules for Filing Catalog Cards* (Chicago: American Library Association, 1968), in which cards are filed in the catalog according to these basic rules:

1) Arrange all entries, English and foreign, alphabetically according to the English alphabet.

2) Arrange word by word, alphabetizing letter by letter to the end of the word. (This is the rule "nothing precedes something"; example: New York precedes Newark.)

3) Items that are disregarded:
    A. The articles *a, an* and *the* in initial positions are disregarded, but when they appear elsewhere, they are given the same treatment accorded any other word. Articles in all languages are treated in the same manner.
    B. Designations such as *comp., ed., illus., jt. author, pseud.*, and *tr.*, when they appear in entries, are disregarded.
    C. Designations such as *Sir* and *Gen.*, when they appear in inverted personal names are disregarded.
    D. Commas, periods, parentheses, apostrophes and other marks of punctuation are disregarded.

4) Abbreviations.
    Arrange abbreviations as if spelled in full.
    Examples: Mc or M' as if Mac; St. as if Saint

5) Elisions.
    Arrange elisions in English as they are printed and not as if spelled in full. Example: "O'mine" *not* "of mine." Treat as one word the contraction of two words resulting from an elision. Example: "Who's" is filed "Whos," *not* "Who is."

6) Arrange numerals in the titles of books as if spelled out in the language of the title. Spell numerals and dates as they are spoken, omitting the "and" except at a decimal point between two digits and in mixed numbers.
   Examples: 101 as one hundred one
   1812 as eighteen twelve, if a date: otherwise as eighteen hundred twelve
   6½ as six and one-half

7) Signs and symbols.
   Alphabetize the ampersand (&) as "and," "et," "und," etc., according to the language used in the title.

8) Hyphenated and compound words.
   Arrange hyphenated words as separate words if each word is a word in itself. If the first part is a prefix such as anti- , co- , etc., arrange as one word.

9) Compound names.
   Arrange names consisting of two or more words, with or without a hyphen, as separate words, after the simple surname, interfiled in alphabetical order with titles and other headings beginning with the same word.
   Examples: Hall, William
   Hall & Patterson
   Hall Family
   Hall of Fame
   Hall-Quest, Alfred
   Hall-Wood, Mary
   Hallam, Arthur

10) Names with a prefix.
    Arrange a name with a prefix as one word. This includes such names as D'Arcy, Du Challu, Van Dyke, Van Loon, etc.

11) Forename entries.
    Arrange a forename entry after the surname entries of the same name, interfiling with titles and other headings beginning with the same word. Include compound forename entries. Alphabetize with regard to all words, articles, and prepositions included.
    Examples: Charles, David
    Charles, William
    Charles-Roux, Francois
    Charles [a title]
    Charles Alexander, duke of Lorraine
    Charles, archduke of Austria
    Charles City, Iowa
    Charles II, duke of Lorraine

12) Author entries.
    A. Under an author's name, personal or corporate, arrange the items in three categories.
       i. Main entries for works by the author, subarranged by title. Literary works may then be subarranged by publisher alphabetically.
       ii. Secondary entries for the author, subarranged by the main entry of the work.
       iii. Works about the author, subarranged by the main entry of the work.

    B.   The entries for two or more persons who have identical names are arranged chronologically by birth date.

13)   Subject entries.
    A.   Arrange a subject, its subdivisions, etc. in the following order:
        i.   Subject without subdivision.
        ii.   Form, subject, and geographical subdivisions, inverted subject headings, subject followed by a parenthetical term, and phrase subject headings interfiled in one alphabet, disregarding punctuation.
        iii.   Period divisions under such subheads as *History, Politics and government*, and *Foreign relations* arranged chronologically.

14)   Order of entries.
When the same word, or combination of words is used as the heading of different kinds of entries, arrange the entries alphabetically by the word following the entry word. Disregard kind of entry and form of heading, except as follows:
    A.   Arrange personal surnames before the other entries beginning with the same word.
    B.   Subject entries under a personal or corporate name are to be filed immediately after the author entries for the same name.
        Examples: Love, John L
                LOVE, JOHN L
                Love.
                  Smith, John.
                Love.
                  Taylor, Robert.
                LOVE
                  Williams, Thomas.
                Love and beauty.
                LOVE (IN THEOLOGY)
                Love-letters.
                A love match.
                LOVE POETRY
                LOVE – QUOTATIONS, MAXIMS, ETC.
                Love songs, old and new.

15)   Editions.
Cards that are the same except for an edition number – e.g., 2nd ed., 3rd ed. – or a notation such as *rev.* are filed in chronological order by publication date.

        Since the filing of catalog cards is a major part of the responsibilities of the technical assistant, the knowledge necessary to complete this task should be both thorough and accurate. Today, the practice of filing above the rod and then checking the cards before dropping them into permanent place has been assigned to the assistant. An excellent teaching unit on filing can be found in Lillian Wehmeyer's *The School Library Volunteer* (Littleton, CO: Libraries Unlimited, 1975).

BASIC SOURCES

American Library Association. *ALA Rules for Filing Catalog Cards*, 2nd ed. abridged. Chicago: ALA, 1968.

Anglo-American Cataloguing Rules, 2nd ed. Chicago: American Library Association, 1978.

Bloomberg, Marty, and G. Edward Evans. *Introduction to Technical Services for Library Technicians*, 4th ed. Littleton, CO: Libraries Unlimited, 1980.

Borkowski, Mildred V. *Library Technical Assistant's Handbook*. Philadelphia, PA: Dorrance, 1975.

Cutter, Charles Ammi. *Cutter-Sanborn Three-Figure Author Table-Swanson-Swift Revision*. distr. Littleton, CO: Libraries Unlimited, 1969.

Dewey, Melvil. *Dewey Decimal Classification and Relative Index*, 19th ed. Lake Placid Club, NY: Forest Press, 1979.

Hicks, Warren B., and Alma M. Tillin. *Developing Multi-Media Libraries*. New York: R. R. Bowker Co., 1970.

National Information Center for Educational Media. *NICEM Indexes*. Los Angeles, CA: NICEM, 1969- .

Prostano, Emanuel T., and Joyce S. Prostano. *The School Library Media Center*. Littleton, CO: Libraries Unlimited, 1971.

Sears, Minnie E. *Sears List of Subject Headings*, 11th ed. Edited by Barbara M. Westby. New York: H. W. Wilson, 1971.

Sive, Mary R. *Selecting Instructional Media: A Guide to Audiovisual and Other Instructional Media Lists*, 2nd ed. Littleton, CO: Libraries Unlimited, 1978.

Tillin, Alma M., and William J. Quinly. *Standards for Cataloging Nonprint Materials*, 4th ed. Washington, DC: Association for Educational Communications and Technology, 1976.

United States Library of Congress. *Audiovisual Materials*. Washington, DC: Library of Congress, 1980.

United States Library of Congress. *National Union Catalog*. Washington, DC: Library of Congress, 1956- .

United States Library of Congress. Subject Cataloging Division. *Library of Congress Subject Headings*, 9th ed. Washington, DC: Library of Congress, 1980. (2 vols.).

Wehmeyer, Lillian B. *The School Library Volunteer*. Littleton, CO: Libraries Unlimited, 1975.

Wynar, Bohdan S. *Introduction to Cataloging and Classification*, 6th ed. Littleton, CO: Libraries Unlimited, 1980.

The Book Trade in Classical Times, ca A.D. 100

4—PHYSICAL PROCESSING AND STORAGE OF AUDIOVISUAL MATERIALS

The variety of shapes and sizes of audiovisual materials makes decisions regarding the physical processing and storage of these materials an important one. The specialist must decide whether to develop an integrated collection (print and nonprint media shelved together) or a separate collection (nonprint media shelved and stored in a location separate from print media), and what procedures for physical processing should be implemented to facilitate ease of access and to ensure effective circulation and utilization.

PROCESSING PROCEDURES

The responsibility of preparing the materials for circulation and access upon completion of cataloging activities by the specialist is generally that of the technical assistant. Simple procedures would be:

1) Transferring materials to the processing area.
2) Placing ownership marks on designated parts of the materials.
3) Typing symbols and call numbers on labels and attaching them to items, or writing them on the containers or items themselves. Adhesive labels of predetermined sizes should be acquired and available (*see* Figure 4-1, page 90).

Figure 4-1. Adhesive Labels

4) Preparing circulation cards for each item, to be kept either with the materials in the containers or at the circulation desk, as determined by the center. The call number, title, and accompanying material designation for each item should be typed on the circulation card and pocket.

5) Pasting the pockets to the containers and inserting the circulation cards in the pockets or filing them at the circulation desk (*see* Figure 4-2).

Figure 4-2. Circulation Card

Filmstrip/Record	FS/R-510-511
KF 4750 .C5	Civil disobedience. 2 filmstrips 2 records with teacher's manual

DATE	ISSUED TO

Filmstrip/Record   FS/R-510-511

KF
4750
.C5

Civil disobedience.
2 filmstrips
2 records
with teacher's manual.

A practice currently in use is to have circulation cards typed and available for audiovisual equipment in circulation. The circulation card is attached to the equipment and kept at the desk when the equipment is in circulation.

CURRENT STORAGE OPTIONS

Current developments in the processing and storage of audiovisual materials have made it possible for collections of such materials to be either totally or partially intershelved. Total intershelving of materials provides for the complete integration of all materials on a shelf. Such arrangement permits a greater and more effective use of all materials by bringing together such material for user browsing and utilization. Partial intershelving provides for the location of materials on individual multimedia shelves or book trucks placed within the regular book shelves. Current shelving may be adjusted to various sizes and for shelving inserts (*see* Figure 4-3).

Figure 4-3. Integrated Storage of Various Materials

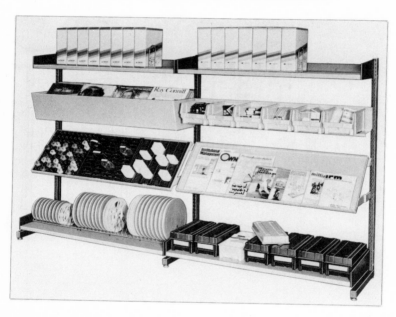

In addition, the variety of audiovisual packaging currently available permits the storage of this material in a boxed book format.

PROCESSING FOR SPECIFIC MEDIA

The following examples demonstrate and illustrate possible processing and storage practices for several major types of materials found in a collection. In processing nonprint media, care should be given to the handling of the materials. Caution should be taken not to bend, scratch, or twist the item in identification, labeling, stamping, perforating (for a leader of film), etc.

Audiodiscs—Phonorecord, record disc, phonodisc, recording disc. Audiodiscs should be handled carefully and by the edges only.

Labeling

Pressure sensitive labels should be adhered to one side of the disc and to the record jacket. The label should be of a size that will not cover vital information on the record and yet be large enough to include the complete call number with media designations, if used, accession number, and the name of the center. The label on the record jacket should be placed in the upper left corner of the jacket front and should also contain the call number, the name of the center, and series data.

Storage

Audiodiscs should be shelved standing on edge either horizontally with the edge out or vertically in a browsing bin as shown in Figure 4-4.

Figure 4-4. Storage of Audiodiscs

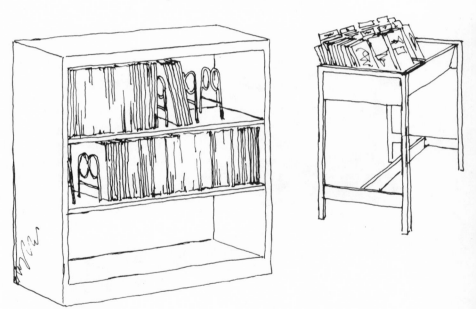

Film Media—Motion pictures, filmstrips, slides. Film should be kept free of dirt and dust by continuous cleaning and inspection after each use. It should be handled by the edges and not bent, twisted, or scratched.

Motion Pictures

Labeling: Leader and trailer should be labeled with the call number, accession number, and the name of the center as shown in Figure 4-5. The reel should be labeled with a pressure sensitive label that includes the call number, accession number, and title (*see* Figure 4-5). The container should also be labeled with the title, call number, accession number, and the name of the center. In addition, the rim of the container should be marked with the call number and/or accession number as shown in Figure 4-6.

Figure 4-5. Leader with Label

```
MP                          MP-75
HN
17.5
.F8       Future shock.


University of D.C.
Library/Media Serv.
```

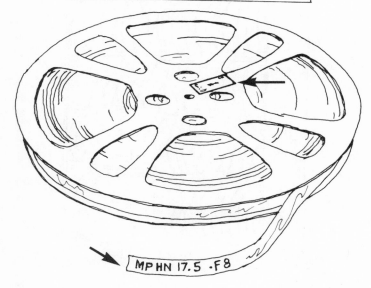

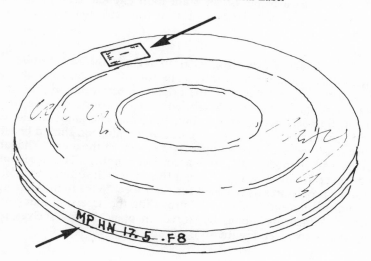

Figure 4-6. Rim of Container with Label

Storage: Motion picture films should be stored on their edges with the call or accession number label on container visible (*see* Figure 4-7).

Figure 4-7. Motion Picture Films on Shelves

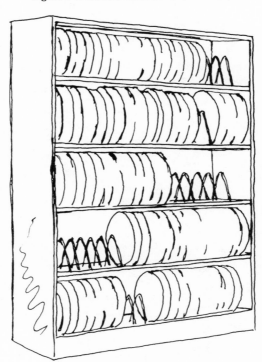

8mm Loops

Labeling: The cartridge of an 8mm loop (standard and Super 8) and its storage container should be labeled with the title, call number, and the name of the center.

Filmstrips

Labeling: Label a filmstrip leader to include the call number, accession number, and the name of the center. The top of the container should also be labeled with the call number with the original information visible. In addition, the side of the container should be labeled with a pressure sensitive label containing the call number, accession number, title, and the name of the center (*see* Figure 4-8). An accompanying study guide or learning aid should be labeled with a pressure sensitive label on the upper right corner of the cover. The label should include the call number, title, and accession number. The accession number should be preceded by "Use with." For a filmstrip with sound (records, cassettes, reels), follow procedures as outlined for each specific medium. Here again, such learning aids should have "Use with" preceding the accession number.

Storage: Filmstrips should be stored on open display shelves for ease of browsing and access (*see* Figure 4-9).

Figure 4-8. Filmstrip and Container with Label

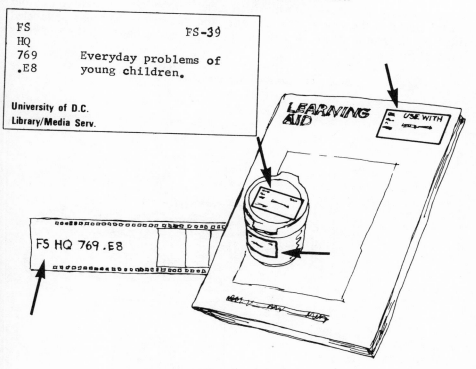

Figure 4-9. Filmstrip Display Shelves

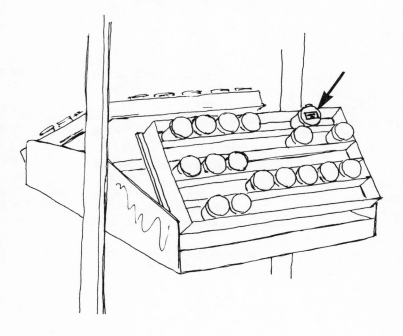

Slides

Labeling: Label each slide with either a call number or an accession number and the name of the center. The slide tray and package should also be labeled with the call number, title, accession number, and the name of the center as shown in Figure 4-10.

Figure 4-10. Slide, Tray and Container with Labels

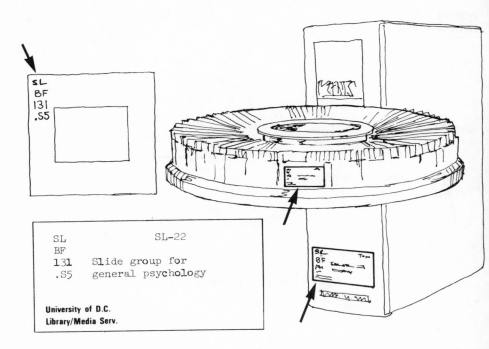

When considering slides with sound (records, cassette, reels), follow procedures as outlined for each specific medium. Here again, such learning aids should have "Use with" preceding the accession number.

Storage: Slides may be stored in sets in slide trays and kept in boxed containers for shelving. Cabinets especially designed for displaying slides are also available.

Microforms (microfilm, microfiche, microopaques, aperture cards)

Labeling: Microforms should be labeled on all individual items and the container. The leader of the microfilm and the microfiche envelope should be labeled with the call number or accession number and the name of the center. The reel, box container, or cartridge should also be labeled with pressure sensitive labels as shown in Figure 4-11.

Storage: Microforms may be packaged in boxes or cartridges, and stored on shelves or in specially designed cabinets (*see* Figure 4-12).

Figure 4-11. Microforms with Labels

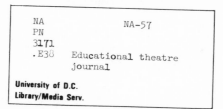

```
NA                    NA-57
PN
3171
.E38    Educational theatre
        journal
```
University of D.C.
Library/Media Serv.

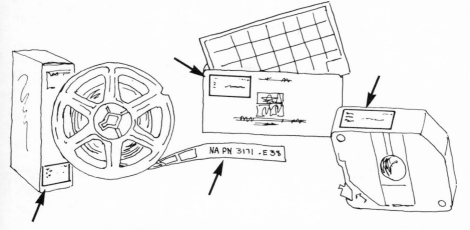

NA PN 3171 .E38

Figure 4-12. Microforms in Storage Cabinets

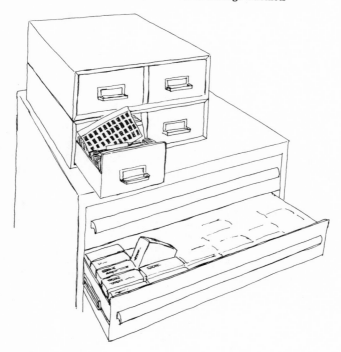

Magnetic Tapes — Audiotapes and videotapes. Magnetic tapes should be handled carefully and stored in dust-proof containers.

Audiotapes — Reel
Labeling: A tape leader should be labeled with the call number and the name of the center. The reel may be labeled with a pressure sensitive label containing the call number. The storage box should be labeled on its edge with the call number and title.

Audiotapes — Cassette
Labeling: The call number and the name of the center should be marked on the original label. The storage box should also be labeled to include the call number, title, accession number, and the name of the center as shown in Figure 4-13.

Figure 4-13. Cassette Tape with Labels

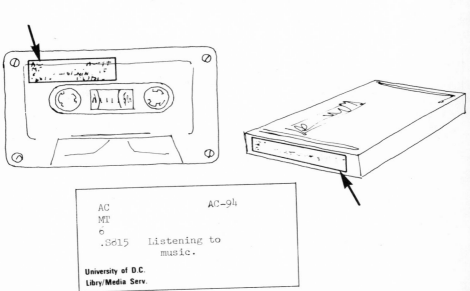

Storage: All audiotapes may be stored in storage boxes on shelves in an upright position with the spine of the box visible (*see* Figure. 4-14).

Videotapes
Labeling: The leader of reel tape should be labeled with the call number, accession number, and the name of the center. The tape box spine should also be labeled. A cassette cartridge should be labeled on the spine with the call number, accession number, title and the name of the center.
Storage: Videotapes may be stored in boxes on shelves with the spine visible. Specially designed tape cabinets are also available for protection from dust and possible damage.

Figure 4-14. Audiotapes on Shelves

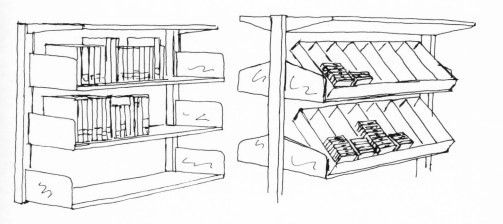

Three-Dimensional Materials—Kits, globes, games, models, realia. Three-dimensional materials include a wide variety of compositions and shapes, and should be shelved with or near other media in the same subject field.

Kits

A kit usually consists of two or more media cataloged as a unit.

Labeling: Label each type of material in the kit as outlined and illustrated in the examples for each in this section. The container should be labeled with a pressure sensitive label to include call number, title, the name of the center, and the accession number. An accompanying teaching or learning guide should also be labeled, and the accession number on the label should be preceded by "Use with." A listing of complete contents should be indicated on the card or borrower's pocket inside the lid container (*see* Figure 4-15, page 100).

Storage: Kits may be stored in containers on the shelves in the same manner as a book. The spine of the container should be visible for easy call number identification.

Globes

A map of the earth on a sphere to give an accurate representation of the earth. A wide diversity of sizes and types of globes are available, and some are permanently mounted; a portable dust cover may be used for protection.

Labeling: Label the pedestal or base of the globe with the call number, accession number, and the name of the center (*see* Figure 4-16, page 100). Accompanying teaching materials should also be labeled, and the "Use with" notation should precede the accession number.

Figure 4-15. Kit with Labels

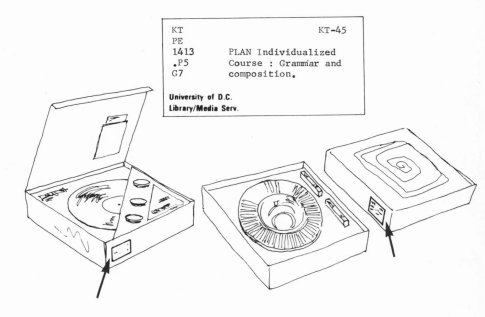

```
KT                              KT-45
PE
1413        PLAN Individualized
.P5         Course : Grammar and
G7          composition.

University of D.C.
Library/Media Serv.
```

Figure 4-16. Globes with Labels

```
QG              QG-2
G
3170    National Geographic
1969       Society.
.N3  National Geographic
          globe.
University of D.C.
Library/Media Serv.
```

Storage: Globes may be stored in cabinets, on carts, or in a permanent location with a dust cover (*see* Figure 4-17).

Figure 4-17. Storage of Globes

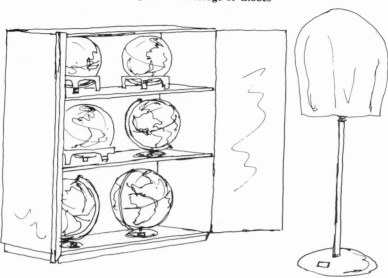

Games

Instructional materials designed to be used to test skills through competition or play.

Labeling: Label the container on the outside edge with the call number, accession number, title, and the name of the center. The inside lid should contain a borrower's pocket, which should include the call number, accession number, title, list of contents, and the name of the center (*see* Figure 4-18).

Figure 4-18. Game with Labels

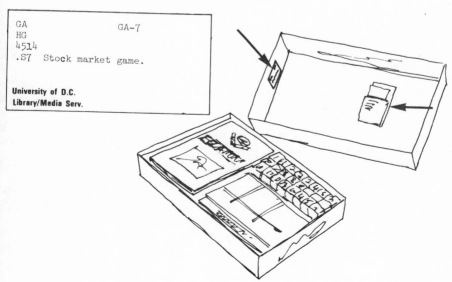

Storage: Games may be shelved in either a flat or upright position with the call number visible on the outside box.

Models

Three-dimensional representation of an object, either exact or to scale, mock-up. Plastic bags or covers should be used for protection.

Labeling: Label the mounting of the model to include the call number, accession number, and the name of the center. Removable parts should also be labeled with the accession number and the name of the center (*see* Figure 4-19).

Figure 4-19. Model with Label

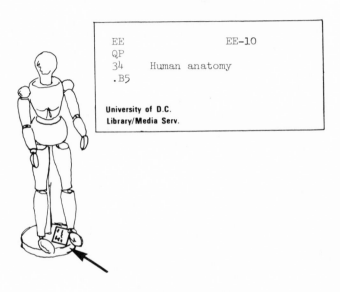

Storage: Models may be stored in cabinets or housed on open shelves.

Realia

Real objects, samples, artifacts, specimens.

Labeling: Label each object in a prominent location, the label should include the call number, accession number, and the name of the center. The container, if appropriate, should also be labeled with a pressure sensitive label that includes the call number, accession number, title of the object, and the name of the center (*see* Figure 4-20).

Storage: Realia may be stored on open shelves, in regular cabinets, or in glass display cabinets. Fragile objects should be housed in cabinets for protection.

Figure 4-20. Realia with Labels

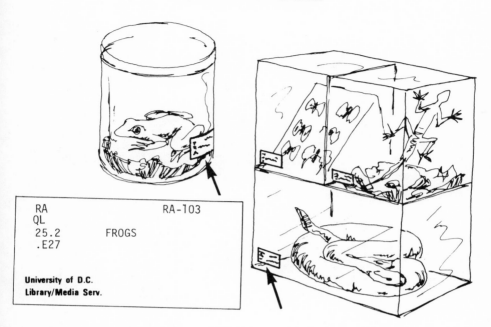

RA
QL   RA-103
25.2   FROGS
.E27

University of D.C.
Library/Media Serv.

Two-Dimensional Opaque Materials—Charts, maps, flash cards, pictures, transparencies, and microfiche. Two-dimensional opaque materials should be handled with great care. Dirt and grease from fingers and other sources can damage these materials. They should be protected through storage in envelopes or by mounting, laminating, edging, spraying, and covering with vinyl picture covers.

Charts

Sheets of information in outline, graph, or tabular form.

Labeling: Label the bottom of the roller or the bottom of the chart. The label should include the call number, title, accession number, and the name of the center.

Storage: Charts may be stored in drawers in a flat position or filed in a filing cabinet much in the same way as vertical file materials. Charts on rollers may be stored in specially designed map or chart cases (*see* Figure 4-21, page 104).

Maps

Flat representations of either the earth or the universe.

Labeling: A map should be labeled on the bottom of the roller or the bottom of the map. The label should include the call number, title, accession number, and the name of the center (*see* Figure 4-22, page 104).

Storage: Maps may be stored in drawers in a flat position or filed in a filing cabinet in a labeled folder(*see* Figure 4-22, page 104). Charts on rollers may be stored in specially designed map or chart cases as shown in Figure 4-21, page 104).

Figure 4-21. Storage of Charts

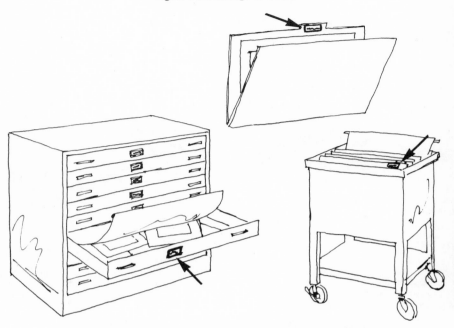

Figure 4-22. Maps, Map Cabinet, and Map Folder with Labels

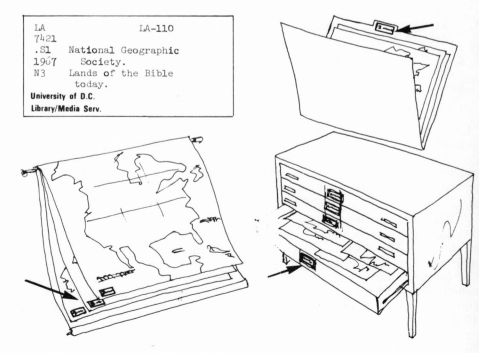

```
LA                LA-110
7421
.S1   National Geographic
1967    Society.
N3    Lands of the Bible
        today.
University of D.C.
Library/Media Serv.
```

Flash Cards

Cards with printed words, pictures, or numerals for rapid identification.

Labeling: Label each card and accompanying material with the accession number and the name of the center. The container should be labeled with pressure sensitive labels that include the call number, title, accession number, and the name of the center. (*see* Figure 4-23). A borrower's pocket should be located inside the lid and should be labeled with the call number, accession number, title, the name of the center, and the number of cards and accompanying materials.

Figure 4-23. Flash Cards with Labels

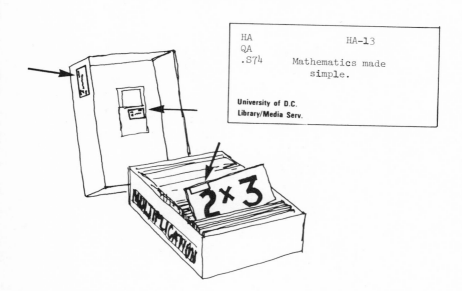

Storage: Flash cards may be stored on open shelves with the call number visible.

Pictures

Pictures, photographs, art prints, art originals, posters, and study prints.

Labeling: Label a picture on the reverse side with the call number, accession number, and the name of the center. The borrower's card should be located on the folder or envelope containing the picture. It should also be labeled with the call number, title (series), accession number, and the name of the center.

Storage: Pictures should be stored in folders or envelopes and filed in much the same way as vertical file materials in a filing cabinet.

Transparencies

Images produced on transparent material.

Labeling: Label each transparency with a pressure sensitive label that includes the call number, accession number, and the name of the center. The labels should be placed on the front side of frame and on the envelope or folder.

Storage: Each transparency may be stored in an envelope or a folder and filed in a cabinet (*see* Figure 4-24, page 106).

Figure 4-24. Storage of Transparencies

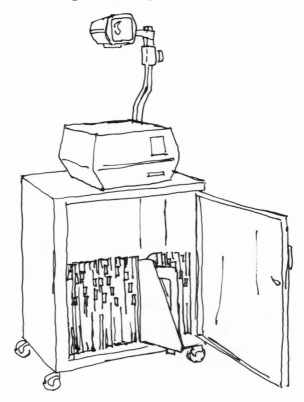

Machine-Readable Data Files (MRDF) – A MRDF is a collection of related records treated as a unit and represented in such a way that it can only be read and/or translated by a machine. MRDFs include files stored on magnetic tape, punched cards with or without a magnetic tape strip, aperture cards, punched paper tapes, disc packs, mark-sensed cards, optical recognition font documents, etc.

Storage

The storage of machine-readable data will vary according to the format of the file. Information stored in one format may be transferred to another, e.g., punched cards to magnetic tapes.

BASIC SOURCES

Hill, Donna. *The Picture File: A Manual and a Curriculum-Related Subject Heading List*, 2nd ed. Hamden, CT: Shoe String, 1978.

Irvine, Betty J., and P. Eileen Fry. *Slide Libraries: A Guide for Academic Institutions, Museums, and Special Collections*, 2nd ed. Littleton, CO: Libraries Unlimited, 1979.

Johnson, Jean Thornton, and others. *AV Cataloging and Processing Simplified.* Raleigh, NC: Audiovisual Catalogers, 1971.

Larsgaard, Mary. *Map Librarianship.* Littleton, CO: Libraries Unlimited, 1978.

Miller, Shirley. *The Vertical File and Its Satellites: A Handbook of Acquisition, Processing and Organization*, 2nd ed. Littleton, CO: Libraries Unlimited, 1979.

Shaffer, Dale E. *Library Picture File: A Complete System of How to Process and Organize.* Salem, OH: D. E. Shaffer, 1972.

Weihs, Jean Riddle, Shirley Lewis, and Janet MacDonald. *Nonbook Materials: The Organization of Integrated Collections*, 2nd ed. Ottawa: Canadian Library Association, 1979.

PART II
Using Audiovisual Materials and Equipment

Stag Heads, Lascaux Caverns, Montignac, France

5 — VISUAL DISPLAYS

The drawings on the walls of the Lascaux Caverns in France, as are those in many other caves found around Europe, are examples of man's earliest uses of graphic symbols. These ancient drawings probably have ritualistic meaning, and are testimony of communication through a graphic form. Such beginnings were followed by centuries of development in which the graphic arts became an important component in many cultures. Alongside the improvement of techniques and methods, design rules and principles were formulated to explain and to further all forms of visual communication. A poster, a photograph, a display, a book page, a motion picture frame will obey the principles of design. The proper use of design and its elements will enhance the message. In your work, you will employ these principles often. You must learn to use the elements of design and to follow its principles in composition and layout.

One of the best ways to start learning composition is by studying objects which are very common — objects we frequently overlook. Billboard displays, advertising posters, magazine photographs, drawings, and paintings are around us everywhere. From the moment we sit at the breakfast table until we return home in the evening, we are confronted with numerous forms of visual displays. Start looking at them. Think about how they attract your attention. Compare different designs and different media.* You will soon begin to realize that each design contains a variety of components or elements, and to recognize how they are employed and what effect they produce on you.

DESIGN AND ITS ELEMENTS

Design is, basically, the arrangement of elements in a manner pleasing to the human eye. These elements are *line, shape, color, texture*, and *space*. Their use, their combination, their placement will create an effect, excite a sensation. They can simulate motion, attract attention, convey an idea, elicit an emotion. By manipulating these elements you will communicate. How well you do it will depend on how well you learn to handle them.

A *line* is technically a series of dots joined together.

It may be long, short, thin, bold, graceful, continuous, broken, straight, or curved.

If a line thickens at one point, a *shape* forms.

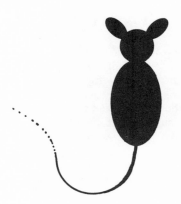

---

*The definition of "medium" as it has been used so far in this book should be expanded at this point. In earlier chapters it has been used principally to designate the technology that serves the dissemination of information. Thus, we have referred to the medium of print as well as to nonprint media, which include filmstrips, films, graphic materials, realia, sound recordings, etc. In art, information (and emotion) is transmitted through the media of paint, stone, steel, plexiglass, ink, wood, any material the artist believes fits the sense of the work.

With the use of dyes, *color* can be added to the original dot. Pencils, ink, paints of different hue and consistency are available.

By applying these media in different amounts, concentrating more paint in one spot than in others, a *texture* develops. Try applying paint with a brush, then with a flat object, even with your finger. Press different objects on fresh paint such as burlap or sand paper. You will learn to understand and to visualize the elements of design and its effects by experimenting.

All the above-mentioned elements occupy *space*. For practical purposes, our space will be a two-dimensional surface, such as a page, a board, a slide. The surface we commonly work on has only two dimensions, width and height, not the third dimension of depth.

PRINCIPLES OF DESIGN

As stated previously, the arrangement of elements on a surface will obey certain principles and have certain characteristics. A good design will be said to have *balance, harmony* or *unity*, and *clarity*.

Balance can be formal or informal. In a formally balanced layout, if an imaginary line divides the whole, each half will be roughly equal to the other half (*see* Figure 5-1). It is a *symmetrical composition*.

Figure 5-1. Symmetrical Composition

Formal balance gives a sense of security and stability, but it is usually static. Informal balance places the object off center in an *asymmetrical composition* (*see* Figure 5-2).

Figure 5-2. Asymmetrical Composition

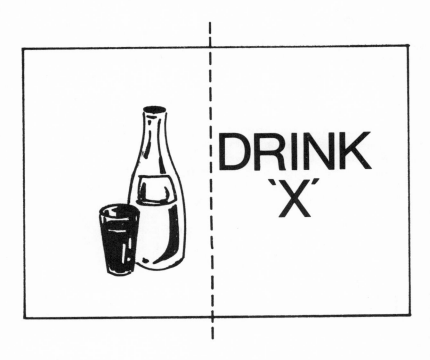

Informal balance is more interesting. It adds excitement to the composition, and it is usually dynamic. It also requires more work and know-how to create, but it is preferred in many cases.

Harmony refers to elements of a layout in agreement with each other. If colors are used, their hues should be of similar tonality. Lines should be of similar thickness and shapes of similar nature. The design should give a sense of *unity*, of all elements belonging together. A simple way of arranging shapes harmoniously on a layout is by following a grouping suggested by a letter of the alphabet which is in itself balanced such as the X, the Y, the I, the Z, the S, etc. (*see* Figure 5-3). The ability to communicate effectively is sometimes synonymous with how simply we express ourselves. In visual display, *clarity* and *simplicity* are very important; a slide, a chart, or a display board should be easy to see, to understand, to grasp visually. Too many words on a slide frame, too many clustered photographs on a board, too many lines on a drawing will hinder understanding and the design will be unsuccessful. A bold, clear, simple design is attractive and interesting. If you have lots of information to display don't try to force everything into a limited space; the amount of information must be adapted to your surface area. Be selective; eliminate; condense.

Figure 5-3. Layout Designs

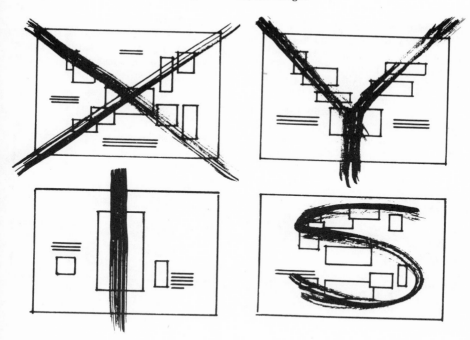

For example, a title slide should not have more than two or three brief lines. Do not write a long sentence for a title. An overhead transparency should not have more than 10 written lines; a movie frame should not have more than two or three lines of written information on it (*see* Figure 5-4).

Figure 5-4. Titling a Display

An Improperly Titled Display

A Properly Titled Display

Judge the size of the surface in relation to the given information. Allow for the elements in your composition to have "room." Don't make your designs as crowded as a subway car in a busy metropolis during rush hour.

Another too often disregarded characteristic in a layout is margins. All good compositions have adequate margins. The margins must maintain a relationship with the size of the design. An 22x14-inch poster may have 3-inch wide margins, while one 4x8 inches may have 2-inch margins. It is also preferable that the bottom margin be wider than the other three. Obviously, in borderless designs there are no margins, but be cautious if words or symbols are to be added. In such cases, observe an imaginary margin (*see* Figure 5-5).

Figure 5-5. Margins on Borderless Layouts

In preparing a design, we strive to present information in a pleasing form, and in most cases we are also interested in focusing attention, in highlighting a particular word, shape, idea, etc. In a word, we want to emphasize *something*. Emphasis can be achieved by applying the elements or the principles we have just mentioned. A line can be used to draw attention to the main point of a composition either by aiming at it or by underscoring it. Shapes can vary in size; a large shape among smaller units will stand up in a layout. Also, the introduction of a different shape to a series of similar shapes will attract the eye.

Contrast, the juxtaposition of light and dark, is another technique to use for emphasis. In a dark background a white spot will catch the eye. A deep, bright color spot on a different color field will achieve the same effect; any note of color on a black-and-white design will be easily seen. Utilizing another form of contrast, texture on a smooth surface can also serve as an attention-getting device. The unusual, unexpected, dissimilar, discordant, bold, will create emphasis in any design.

LETTERING

It is fair to assume that most visual displays will have a message and the the pictorial message may be accompanied by written information. A title, a caption, a descriptive paragraph requires the use of letters, words, sentences. With the exception of speech, writing is still, even in our technologically oriented world, the most commonly found means of human communication. We often tend to overlook the written message, but good lettering can enhance a visual presentation just as much as it can destroy it. Lettering refers here to any and all forms of the written alphabet as well as to the action of writing. To add words to your visuals you can use the following methods:

1) Freehand lettering
2) Stencils or templates
3) Dry-transfer materials
4) Precut letters

Freehand Lettering

This method can be used for informal lettering. You will achieve good results with a little practice. Use some inexpensive paper, such as newsprint or brown wrapping paper, to practice. The hand and the arm should move together when lettering; this means that the wrist should not flex. The arm pivots at the elbow and shoulder joints only. At the beginning, try making large characters; this makes it easier to control and coordinate the arm movements. Draw parallel lines in the working surface to aid your efforts. Initially, you may wish to sketch the letters faintly with a pencil to guide your lettering style, height, and spacing. Make letters with a full stroke; don't stop and start during a continuous line. Once you are satisfied with the results, try the method on the final surface.

The quality of the paper or illustration board to be used should be selected carefully. Since most of the media used in freehand lettering are liquid, an absorbent paper will make the inks or paints spread. This spreading or *bleeding* can be minimized by using specially prepared surfaces. They will accept inks well and your letters will have a sharp, clear *contour*. A visit to an art supply store will prove beneficial. A catalog from a large store will often have information about papers and other supplies.

In freehand lettering three basic instruments can be used, pencil, pen, or brush. Pencils are easy to use. They make a sharp line and are not messy. Pencil lines can usually be erased and come in a large number of colors. Pencils are good for drawing thin lines, but trying to cover a large area with pencil results in an uneven coating. It is difficult to spread the pigment evenly. But a variety of color pencils will, after applying the pigment, blend like watercolor when the area is wetted with a damp dabber. This is a useful technique for some applications, but the colors are usually not intense. It is also a time-consuming practice. To cover large areas, felt tip markers can be more effective for our purposes. Felt tip markers are found in a large array of colors and in different point sizes and shapes (*see* Figure 5-6, page 118). Of course the markings cannot be erased, and they bleed easily.

In using felt tip markers the stroke should not be repeated because this will usually result in different densities of the ink deposit due to variations in the hand pressure. Felt tip markers are found with water soluble and permanent inks. If a water soluble ink marker has dried because of extended storage, it can be "rejuvenated" by dipping the point in water for a while. The water is absorbed through the felt tip, the pigment is dissolved, and the marker may write again.

Using pen and ink for freehand lettering requires practice and care due to the possibility of ink spilling or dripping. *Quills* or *nibs* are made for left-handed and right-handed artists and are found in many sizes and shapes. Their selection determines the letter style and the width of the stroke. Metal quills provide a sharp line and adapt to a variety of lettering styles. The older pen holder and nib, which had to be dipped in ink, have for the most part been replaced by a *reservoir*

Figure 5-6. Several Shapes of Felt Marker Tips

pen, in which an "ink well" is part of the holder. This type of pen minimizes spilling or spotting. There are several varieties of pens and a large selection of interchangeable nibs for these pens are available (*see* Figure 5-7).

Figure 5-7. Fountain Pens and Nibs

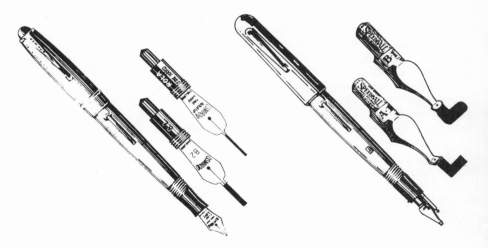

A variation of a reservoir pen is the Rapidograph®. This pen has a tubular nib. The flow of the ink is regulated by a plunger on the nib. When the nib rests on the writing surface, the plunger is raised and the ink flows by gravity. When the pen is lifted, the plunger falls closing the nib opening (*see* Figure 5-8). Due to its nib design, which permits exact control of the line thickness, the pen must be used vertically and is used mainly for mechanical drafting (*see* Figure 5-8).

Figure 5-8. A Rapidograph® Pen

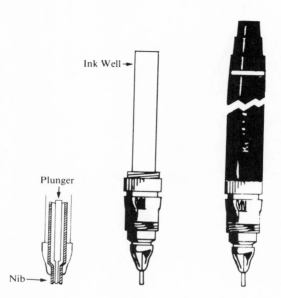

Ink Well →

Plunger

Nib →

The use of brushes in lettering is a more advanced technique, which requires much training and should be left for the experienced graphic artist.

Stencils and Templates

Templates and stencils have the advantage of providing a uniform letter size and style regardless of the lettering skill of the user. They are made of paper, cardboard, wood, plastic, metal, etc., depending on their application and the medium used with them. Stencils and templates are made in various letter styles and sizes and for different shapes and symbols. The words *stencil* and *template* are used interchangeably at times. Stencils are made of thin sheets with dots, lines, or shapes cut out (*see* Figure 5-9, page 120). Through these openings, ink or paint is applied to the surface beneath the stencil through the stencil face. A template is similar to a stencil, except that it is usually a guide for a writing instrument such as a pen or pencil (*see* Figure 5-10, page 120).

The most common and inexpensive stencils are made of thin, waxed cardboard and are used mainly with spray paints. Stencils used for spraying have one character in each, to allow for an entire word or line to be set up and sprayed in one pass (*see* Figure 5-11, page 121). After some use they can be discarded and replaced at low cost. When much use is made of stenciling with paints, metal will replace the cardboard. Paint can be washed off the metal easily, and the stencil can be reused. Templates are more frequently used with pencils and Rapidograph®-type pens (*see* Figure 5-12, page 121).

Figure 5-9. Stencil

Figure 5-10. Template

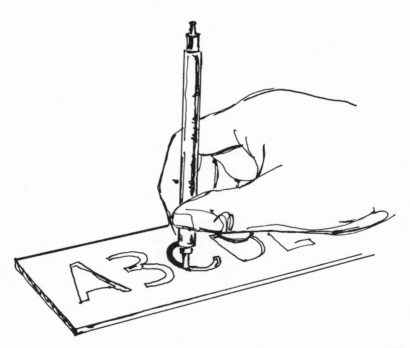

When using ink with a template, care should be given not to have the template resting directly on the working surface. There should be a space between the tracing edge of the template and the ink flowing from the pen. As illustrated in Figure 5-12, many templates have a beveled or undercut edge to prevent blotching. Without this precaution, the ink will seep between the template and the writing surface, blotching the work. If the template you are using does not have this feature, blotching can be avoided easily by taping a few pennies on the

Figure 5-11. Spraying Stencils

Figure 5-12. Side View of Templates

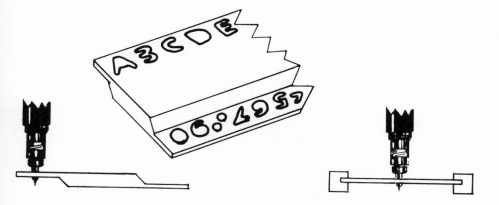

underside of the template. The pennies will raise the template above the working surface and prevent messy results (*see* Figure 5-13, page 122).

Similar spoilage may result when applying paint to a stencil in great quantities. If the paint coverage is not adequate, paint should be allowed to dry before applying a second coat. Also, wait until the paint is dried before removing the stencil, or fresh paint accumulated on the stencil edges may spoil the work.

Several kinds of mechanical tracing systems use a slightly different type of template. With these systems the writing instrument does not pierce the template, but rather the template has grooves that guide a *scriber* or needle attached to an ink well with a Rapidograph®-like nib on it. As the scriber is guided along the character groove the nib will letter on the writing surface (*see* Figure 5-14, page 122). A large selection of templates with different letter styles and sizes are available with this system.

Figure 5-13. Template Raised with Coins

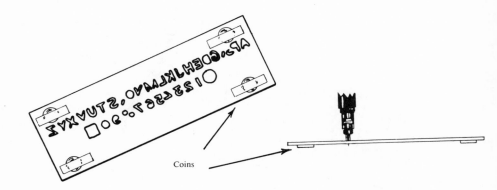

Coins

Figure 5-14. A Leroy Scriber

Another system, the Varigraph®, employs a metal template of high precision, which is inserted into a guiding mechanism, and a Rapidograph®-like pen (see Figure 5-15). As the figures below illustrate, a single alphabet template allows for a variety of letter heights, widths, and slants (see Figure 5-16). To develop skill in the use of these instruments requires a little practice, but the results are well worth the efforts.

Figure 5-15. Varigraph®

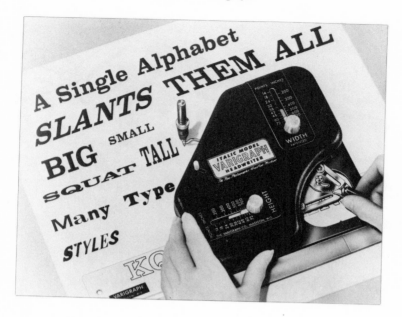

Figure 5-16. Varigraph® Letter Variations

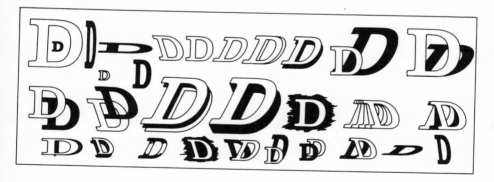

A different lettering technique could be considered a "stencil in reverse." Rubber letters are laid on the working surface and then paint is sprayed over the whole. Once the paint has dried the letters are removed. The places covered by the letters have not received any coloring and the characters will show with the color of the working surface (*see* Figure 5-17, page 124). With this technique, spraying is done in such a way that all sides of the letters receive paint at the same angle to prevent fuzzy, undefined contours. Observe in Figure 5-18 (page 124) how spraying should be done to obtain clean-cut edges. Spraying from the left alone would cause a fuzzy edge on the right of the character, so spraying should be done from *all* sides equally. To obtain an even coating on the working surface spray lightly, then allow to dry, and if necessary spray again. Too much paint at once may run

Figure 5-17. Rubber Letters

Figure 5-18. Spraying 3D Letters

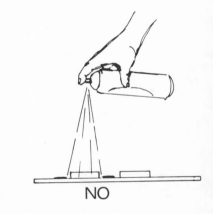

NO

YES

down the side of the letters and drip on the work. Again, allow the paint to dry before removing the letters.

In freehand lettering, we usually work with a wet medium, such as ink or paint. The lettering, although acceptable in many applications, is mostly informal. Unless, we become masters in the art of calligraphy, freehand lettering will be used for display captions, temporary signs, or the like. As soon as we look for lettering to be used in printed materials, slides, film titles, etc., where a more formal and precise character is required, we must use methods even more precise than the template lettering systems previously described.

Dry Transfer

Dry transfer is, without question, the most versatile lettering medium presently available to anyone without previous training in the graphic arts and with no access to typesetting equipment. Dry transfer materials offer the largest number of lettering styles and sizes in most known written languages. They offer unparalleled character definition. Letters can be photographically blown up in size greatly before revealing irregularities or fuzziness at their edges. Dry transfer letters also are available in a wide variety of sizes, colors, symbols, and styles for any layout. The design possibilities are countless. Dry transfer requires very little manual dexterity, though it does require patient and careful handling. These materials require no special equipment of any sort and can be applied to almost any smooth, clean surface.

Dry transfer characters are printed on a transparent sheet. The printed characters have an adhesive coating on the exposed side. The sheet is placed on the working surface with the adhesive side down, and the character is rubbed or *burnished* with a blunt instrument. When the sheet is then lifted the letter remains on the working surface (*see* Figure 5-19, page 126). This simple method requires some consideration in the transferring steps. On the working surface, faint pencil guide lines should be drawn. Since the sheet printed with the transfer symbols is semi-transparent, this will allow you to align the letters with one another. After the symbol or letter is in the desired position, it can be burnished. The burnisher can be either a specially designed tool or a soft pencil with a blunt point. Care should be taken not to move the sheet while rubbing; moving the sheet will crack the letter. The entire letter must be burnished with an even pressure. The letter will look lighter in color as you burnish it off. The milky-grey color of the carrier sheet will become visible. When the whole letter is burnished the carrier sheet should be lifted as if peeling off a label. Don't jar the sheet aside. A peeling-off motion will leave the symbol cleanly behind on the working surface, often even if the burnishing was not fully completed.

Most dry transfer sheets come with a loose backing that protects the exposed adhesive side of the characters during shifting and handling. This very thin sheet is usually coated with silicone; no adhesive will stick to silicone. Once the transfer is finished, cover the work with this sheet and rub firmly with the burnisher over the transferred characters. This will ensure good adhesion. After this step, the guide lines can be erased, but be careful not to go over the symbols harshly. In cases where the working surface prevents drawing a guide line or as an alternative to this practice, the edge of the protective sheet can be used as a guide. Simply attach the sheet to the drawing board or the working surface with one of its edges marking where the letters should be placed (*see* Figure 5-20, page 127).

Dry transfer also allows for the easy correction of mistakes. The characters can be "lifted off" with masking tape or other low-tack tape. Even if the removal must be done after the final burnishing step and the bond is too strong for the tape to break, a cotton swab moist with rubber cement thinner (a solvent) will dissolve the symbol. Wet the symbol for a few seconds and lightly wipe it off with the cotton swab. The symbol will come loose and break up. Allow the solvent to

Figure 5-19. Steps in Dry Transfer

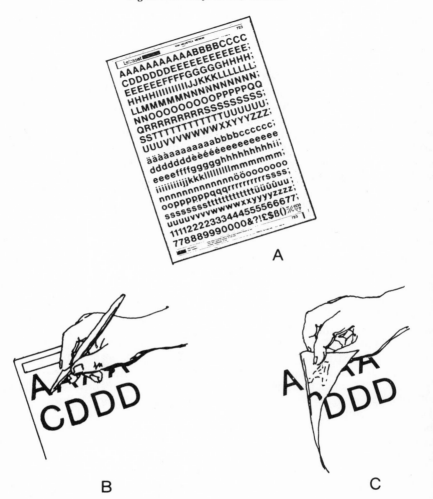

dry completely, before reapplying a character to the same spot. This solvent is very volatile, and it will dry up in a matter of seconds. Dry transfer has few drawbacks, but two things should be mentioned. First, dry transfer materials should be stored under good conditions at average room temperature; if they are not subjected to extreme cold or heat, they will last well over a year. Second, avoid using them in a hot, humid environment, or handling them with warm or wet hands. Since all symbols are exposed on the entire sheet and will come in contact with the working surface while transfering, heat will increase the adhesive action and symbols may come loose without burnishing. After trying the process once you will become adept at it, such are the easyness of application, the professional look of the work, the variety of styles and sizes, and the innumerable opportunities for design combinations and creativity offered by this medium.

Figure 5-20. Using the Protective Sheet as a Guide

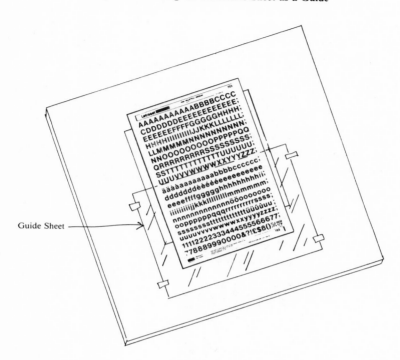

Guide Sheet

Precut Letters

Another alternative for adding captions to your displays is the use of precut letters. There are a great variety of these letters on the market; they are available in many styles, sizes, and materials. These letters can also be obtained with gummed backing, like postage stamps. They need to be moistened to apply. This type is usually found cut out of thin paper, which causes them to curl if too much moisture is applied. Paper letters are the least expensive of this group. The most expensive are cut from thin vinyl and have a self-adhesive backing.

Other types of precut letters are designed for use on cork boards or magnetic panels. Letters to use on cork boards have a set of pins on the back and can be pushed into the board. They are made of clay (plaster) and are very fragile; care should be taken when pinning them. Pressure should be applied evenly, especially on letters such as the R or the K, which may break off at the joints. Magnetic rubber letters are used on metal surfaces. They are available in different colors and are useful for quick lettering used in photographic media. Other letters may be made by using equipment employing photographic techniques or by cutting plastic films. There are several such machines in the market, but they are not yet commonly found, therefore their use is not explained here.

Letter Style

Over the years our alphabet has inspired many calligraphers and artists to create many different letter styles (*see* Figure 5-21, page 128). We stated earlier that good lettering can enhance a design; now we must add that the style of the letters is also very important to our designs and compositions. We have developed certain reactions to the characteristics of the written form; we even associate certain feelings with certain styles. A thin, graceful, ornate lettering style will support display depicting joy or happiness. Such classic style as Old Roman would be a good choice in writing a quote from a Greek philosopher. On the other hand, lettering resembling a free hand brush stroke will not be adequate for a formal quote or an illustration relating to the Edwardian period. Nor should the style known as Gothic be employed to copy the Declaration of Independence. A bold, daring design will benefit little from a thin, gracious, fragile letter style. Equally, a design with a preponderance of square shapes should not incorporate a flowing, round lettering line. Written forms are elements of design, and just as any other element they must be in harmony with the whole. Lettering will add much to a design if the letter style is in tune with the rest of the composition.

Figure 5-21. Basic Letter Styles

The styles available to us are numerous. One leading manufacturer of dry transfer lettering advertises 350 different styles. Catalogs from suppliers and manufacturers of dry transfer materials are a good source for the selection of letter styles. The reader interested in an in-depth coverage of letter styles and related topics is urged to consult Donald M. Anderson's *The Art of Written Forms: The Theory and Practice of Calligraphy* (New York: Holt, Reinhart, and Winston, 1969). This excellent book covers the history of writing from its dawn to the present. It is profusely illustrated with letter styles, designs, etc. Chapters 10, 11, and 12 will give a good idea of how much can be communicated visually with alphanumeric characters. The book also has a very good and extensive bibliography that is partially annotated.

Size

Much of the terminology related to lettering stems from typography, or writing with print type. The styles of type are known as *typefaces*. The size of letters is measured in points. A point is 1/72 of an inch, which means that a letter 1 inch high is said to be 72 points. The average newspaper print is 8 or 10 points high. This book is set in 10 point type. When requesting dry transfer materials or a printing job you will have to identify the letter size in points. Dry transfer and printers' catalogs and type samplers usually illustrate typefaces in available point sizes.

Spacing

There are two methods in spacing letters: mechanical and optical. Mechanical spacing sets the letters at equal distances from each other; the various widths of the letters are not considered (*see* Figure 5-22).

Figure 5-22. Mechanical Spacing

But the preferred method is optical spacing in which the shape and width of the letter (and of those on either side of the letter) determine the spacing (*see* Figure 5-23). Optical spacing aids reading; the pattern formed by the letters and words is more pleasing to the eye.

Figure 5-23. Optical Spacing

To place a sentence or a word in a page we can employ a method that may not in all instances be exact but that is usually adequate. Count the letters in the word or sentence. Count the spaces between words as letters. Now begin your layout by centering the middle letter and begin writing from the center out.

REPRODUCTION METHODS

The lettering methods discussed above have one thing in common: they provide one unit per effort. This means that if you need to duplicate a design you must duplicate the effort as well. In most cases this is not a disadvantage since many layouts or displays are not duplicated. In other instances, informational flyers, class handouts, announcements, posters, directional signs may need to be produced in quantity. Several hundred copies may be necessary for distribution. To letter the same information repeatedly is not only tedious but also inefficient. When multiple copies are desired, we must use a technique that can afford us duplication.

Letterpress Printing

Johannes Guttenberg of Mainz, Germany, is credited with developing the process of printing with movable type. We can now use this invention, which revolutionized the modern world, for producing multiple copies of our work. In printing, movable type letters carved in wood or in metal blocks (*see* Figure 5-24A), are aligned on the flat bed of a printing press. These letters, set in a mirror image of the desired word or sentence, are inked with a roller (*see* Figure 5-24B). The paper or cardboard that is to receive the impression is layed on top of the type and a pressure roller is passed over the paper or cardboard (*see* Figure 5-24C). The result is lettering that may be duplicated quickly and inexpensively as many times as needed (*see* Figure 5-24D).

Obviously, printing is an art considerably more exact and complicated than it is described here, and the variety and complexity of printing presses are great. But for our purposes, we can use this kind of press, called a flat bed letterpress or more often a *sign press*, without becoming expert printers. A sign press is found in most graphic shops and is simple and quick to use. The major effort is to clean up the printing ink from the equipment after the job is completed, but this is a small fee to pay for good, even lettering, produced quickly and in quantity.

Offset Lithography Printing

Large commercial outfits, as well as small printing shops in business, industry, or educational institutions, employ offset lithography printing techniques. Offset lithography is a process derived from lithography, a printing method discovered by Alois Senefelder in Germany around 1800. It is based on the natural antipathy of water and grease. Senefelder discovered that ink applied to a wet lithostone (a fine grained limestone) that had been marked with a grease substance (soap, tallow) would stick only to the greasy characters. The ink would then be transferred to paper by the use of a press that would slide the stone covered by the paper under the pressure of a roller.

Lithography printing has evolved into present-day offset lithography printing, which still uses the grease-repels-water principle. However, the printing surface is now a thin metal plate wraped around a cylinder. This cylinder rotates,

Figure 5-24. Printing Steps

A
Type

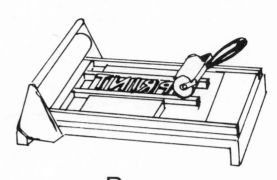

B
Inking

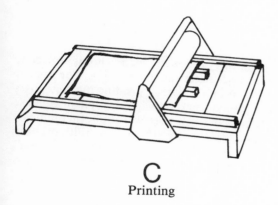

C
Printing

D
Finish Print

coming in contact with water and ink rollers alternately. The cylinder is then pressed against a second cylinder, which is covered with a rubber blanket, transferring or *offsetting* the image onto the blanket. This *offset cylinder* prints on the paper the image picked up from the lithographic plate. The elasticity of the blanket provides a very good printing surface that will reproduce accurate details even on rough or textured surfaces (*see* Figure 5-25).

Figure 5-25. Offset Printing

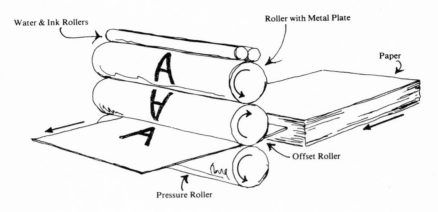

The metal lithographic plates used in this process are made photographically and are usually employed for long runs and better quality printing. For smaller and less expensive jobs, a multilith press, a smaller offset printing machine, uses thinner metal foils or paper plates that can be marked by hand employing grease pencils, india ink, or a typewriter. These plates will also accept photographic or electrostatic reproduction of photographs or type, which makes multilith printing a very attractive method for small shops in schools or universities where small runs of good quality coupled with low costs and relatively easy production steps are of primary concern.

Duplicating

Other equipment and processes are available for making multiple copies on paper, especially for handouts. These are spirit duplicating, mimeograph, thermal duplicating, and several instant duplicating methods.

Spirit Duplication

This is probably the most readily available duplicating equipment. The method employs a drum filled with a liquid chemical or "spirit," usually methyl alcohol, that dampens a sheet of paper, which is then brought into contact with a carbon master wrapped around a drum (*see* Figure 5-26). Just as in a printing press, the paper receives the impression of the information written on the master. This carbon master comes in the usual page sizes of 8½x11 inches or 11x14 inches. Each master package has two sheets: the master and the carbon.

Figure 5-26. Diagram of a Spirit Duplicator

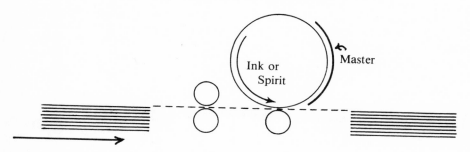

Sometimes a backing sheet is included. If it is not, a sheet of paper should be used to cushion the carbon when writing or typing on the master.

The drawing, lettering, or typing is done on the top or "master" sheet. Mistakes can be corrected on the underside of the master sheet by scratching the surface slightly or applying thin cellophane tape to the spot. Once the master is finished, it is detached from the other sheet and placed on the spirit duplicator. Operating this machine is simple, and the instructions are usually written on the machine itself. A good master can provide approximately 100 good copies. This method usually gives a faint impression, but it is adequate for a short run and when neither quality nor quantity is desired. It is also quick, inexpensive, and easy to use. Care should be given not to let the spirit come in contact with your skin or to breath its vapors. Use in well-ventilated rooms and don't use the spirit to clean your skin.

Mimeograph

In larger institutions stencil duplicating or mimeographing is usually done by duplication or mimeograph units. This method requires a little more practice in its handling than does spirit duplication. This machine forces ink through a drum roller and a plastic stencil to produce an impression on paper or thin cardboard. The preparation of the stencil is similar to that of the spirit master, although a stencil stylus is needed for writing or drawing. The stencil is "cut" well with a typewriter. Mistakes are "erased" by using a special correcting fluid. The fluid will melt the stencil surface slightly, sealing the character or mark. After the spot has dried, the new information can be rewritten on it. This method produces a more intense printing than spirit masters, and hundreds of copies can be obtained from the same master. The stencil can be stored between two pieces of newspaper and used in later runs.

Some thermographic or infrared machines can produce spirit masters and mimeograph stencils from an original that has been drawn or printed with a carbon-based medium, such as common (graphite) pencils, printing inks, or carbon typewriter ribbons. To produce these stencils, the original page is usually inserted in a "sandwich" of the materials required by the individual system and fed into the duplicating unit. The master stencil or mimeo is then handled just as mentioned previously. The thermal unit commonly found in media centers serves this purpose and also duplicates acetate materials. This unit will be discussed with more detail in chapter 7.

Photocopying

In most libraries or offices, it is common to find equipment that will reproduce pages very quickly. These "instant" duplicators or photocopying machines are usually installed and serviced by the manufacturer. The users need only load them with paper and press a button to operate. Copiers made by Xerox®, Canon®, and Kodak® are examples of this type of duplicating equipment. Some can even be used to make transparencies.

MULTIMEDIA INSTRUCTIONAL MATERIALS

Composition. Educational Filmstrips, 1972. 35mm filmstrip, color.

Composition. Educational Media, 1967. 35mm filmstrip, sound, color.

Duplicating by the Spirit Method. BFA Educational Media, 1961. 16mm film, 14 min., sound.

Fundamentals of Layout Design. BFA Educational Media, 1967. 35mm filmstrip, sound, color.

Introduction to Drawing Materials. Film Associates, 1966. 16mm film, 19 min., sound, color.

Introduction to Graphic Design. BFA Educational Media, 1967. 2 35mm filmstrips, sound, color.

Lettering. Doubleday Multimedia, 1970. 35min. filmstrip, sound, color.

Lettering Instructional Materials. Indiana University, 1955. 16mm, 20 min., sound, color or B&W.

Mimeographing Techniques. BFA Educational Media, 1958. 16mm film, 16 min., sound, color.

Mounting Pictures. University of Texas, 1957. 35mm filmstrip, color.

Overhead Transparencies. McGraw Hill, 1968. A series of single concept 8mm film books, silent, B&W.

Signs, Symbols, and Signals. AIMS, 1969. 16mm film, 16 min., color.

Techniques of Modern Offset. A. B. Dick. 35mm filmstrip, sound, color.

BASIC SOURCES

Anderson, Donald M. *The Art of Written Forms: The Theory and Practice of Calligraphy.* New York: Holt, Reinhart, and Winston, 1969.

Arnheim, Rudolph. *Art and Visual Perception.* Berkeley, CA: University of California Press, 1966.

Minor, E. D., and Harvey R. Frye. *Techniques for Producing Visual Instructional Media*, 2nd ed. New York: McGraw Hill, 1977.

Wills, F. H. *Fundamentals of Layout.* New York: Dover Publications, 1971.

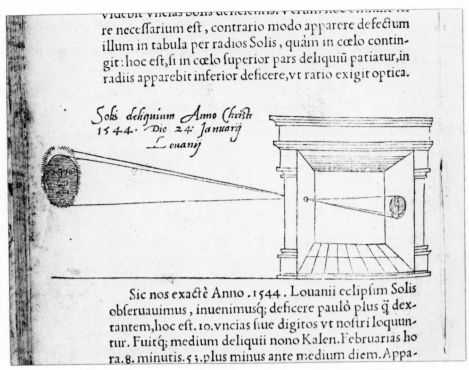

re necessarium est, contrario modo apparere defectum illum in tabula per radios Solis, quàm in cœlo contingit: hoc est, si in cœlo superior pars deliquiū patiatur, in radiis apparebit inferior deficere, vt ratio exigit optica.

Solis deliquium Anno Christi 1544. Die 24. Januarij Louanij

Sic nos exactè Anno .1544. Louanii eclipsim Solis obseruauimus, inuenimusq; deficere paulò plus q̄ dextantem, hoc est. 10. vncias siue digitos vt nostri loquuntur. Fuitq; medium deliquii nono Kalen. Februarias hora. 8. minutis. 5 3. plus minus ante medium diem. Appa-

Gemma-Frisius, First Published Camera Obscura, 1544
(Reproduced courtesy of the Gernsheim Collection, Humanities Research Center,
The University of Texas at Austin)

6 – STILL PHOTOGRAPHY

THE CAMERA

The centuries-old discovery, the *camera obscura*, can be said to be the predecessor of the present day photographic camera. The *camera obscura*, or dark room, phenomenon was discovered by the Greeks centuries before Christ. They observed that when light entered a dark room through a small hole, an image of the exterior surroundings facing the hole was cast, upside-down, on the wall opposite the hole. This phenomenon was later used by Renaissance painters to aid in drawing, and its use spread. During those years, the *camera obscura* began its transformation into a portable instrument, which was later equipped with a lens fitted over the opening to increase the sharpness of the cast image. Basically, this is our contemporary camera: a box in which, through a lens, light enters to expose a light-sensitive material, the photographic film. To this rudimentary camera, other components have been added to perfect its use.

CONTROLLING EXPOSURE

Light entering the camera must be regulated to permit control of the film exposure. The amount of light can be controlled 1) by changing the size of the lens opening or aperture and 2) by regulating the length of time the light is allowed to reach the film.

Lens Aperture

Just as a faucet allows more or less water to flow out of a pipe, the camera lens is equipped with a diaphragm that allows more or less light to enter the camera. This diaphragm or *iris* is made of thin, interlocking metal leaves or blades. They are attached to an outer ring that permits them to close or open and thus change the lens aperture.

The apertures of a camera lens, or *f*-stops, as they are named, are calibrated by a universal standard. The *f*-stops are usually calibrated 1.4, 2, 2.8, 4, 5.6, 8, 11, 16, 22. The *f*-stop markings on the control ring indicate how much light goes through the lens. You need to remember that the *f*-stop numbers are inversely proportional to the size of the aperture opening. An *f*/2 stop represents a larger aperture than an *f*/4 stop. Conversely, an *f*/22 stop is a smaller aperture than an *f*/16 stop. Remember, the larger the number the smaller the aperture (*see* Figure 6-1). The smaller the aperture the less light that reaches the film.

Figure 6-1. Shutter Openings for Different *f*-stops

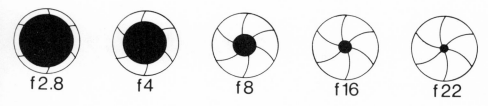

f 2.8   f 4   f 8   f 16   f 22

Length of Exposure

Food left over a fire for an extended period of time burns. If photographic film is exposed to light uncontrolled, it will "burn" or overexpose. Not enough "cooking" time, will produce the contrary effect, underexposure. The span of time that light is allowed to reach the film must be precisely controlled. A shutter placed in the light path blocks the light from reaching the film. A shutter release on the camera allows for this shutter to open, remain open for the required exposure time, and then close.

In the dawn of photography, the first film emulsions needed several hours of exposure before producing an image. Today's emulsions react to light extremely quickly; only an instant is necessary. Shutter speeds indicate how long the shutter stays open. They are marked in fractions of a second: 1, 2, 4, 8, 15, 30, 60, 125, 250, 500, and 1,000. 30 means 1/30 of a second, 125 means 1/125 of a second, and so forth. The shutter can be placed on the lens or in the camera body. In the second case it is like a curtain dropping in and out of the light path.

You can think of the shutter action as the blinking of an eye. In addition to stopping the light from entering the camera, the shutter can help to "freeze" motion. Look in the direction of a person running, close your eyes, and then quickly blink an eye. You will see the person only in one position. You will not see the continuous flow of the action. You have "stopped" the action, isolated one position from a succession of moves. The shutter will duplicate the eyelid function—even better: it can be faster. A fast shutter speed allows this "stopped" action to be recorded on film. The faster the action is, the faster the shutter speed needed to "freeze" it. A racing horse or car may require a shutter speed of 1/250 of a second or even 1/1000 of a second. The same concept also helps minimize camera shakes, which can blur the picture because a given point of light will strike more than a single point on the film. Just as the action can move in relation to the camera, the result on a blurred picture, the camera can move in relation to the action with the same result. A shutter speed of 1/60, or faster will prevent blurring from camera shakes. When slower shutter speed is needed, a tripod to support the camera steadily must be used.

Shutter Speed and *f*-stop

It must have occurred to you already that these two variables interact with each other. Each lens *f*-stop is calibrated to allow either twice as much or half as much light through it. For example, an *f*/8 stop will let twice the amount of light enter the camera as will an *f*/11 stop. Conversely, *f*/11 allows half the light to enter that *f*/8 does; by the same token, a shutter speed of 1/30 of a second will keep the shutter open twice as long as one of 1/60 of a second. This relationship between shutter speed and lens aperture provides us with a series of equal alternatives for our film exposures. As an example, if the correct exposure is to be *f*/11 at 1/60, then the following combinations are equally good:

Shutter Speed =	1/500	1/250	1/125	1/60	1/30	1/15
f-stops =	4	5.6	8	11	16	22

Determining the Correct Exposure

Light sources, as well as their intensity, vary; even natural light changes in strength. To expose film properly we must be able to find out how much light is available. Most cameras have built-in provisions to aid you in exposing the film correctly. In some cameras, a light or exposure meter that reads the intensity of light will allow you to set the shutter speed and the *f*-stop accurately. In other cases, the camera may be equipped with an automatic exposure control mechanisms. Just as your eyes adjust to the available light, a camera with an "electronic eye" will set the appropriate shutter speed and lens aperture for the correct exposure. If your camera has neither of those features, an external light meter is necessary to measure the light intensity. The use of such an instrument will eliminate the wasteful practice of finding the correct exposure by trial and error.

Film

Photographic film has evolved greatly since the first experiments carried out by J. N. Niepce and L. J. Daguerre in the 1820s. Many film types and makes are on the market today. The photographer must know and understand film characteristics in order to select the appropriate film for each job. Film can be identified in two groups, *negative* and *reversal*. When exposed to light, negative film will give a photographic image the opposite of what the human eye sees; that is, the negative will have black values where the eye sees whites, and whites where the eye sees blacks. Shades of grays will also be in contraposition with the "real" values. In the case of color negative film, the process is more complicated. The primary colors will appear as their corresponding complimentary hues. The popular photograph is the positive image resulting from photographically printing the negative (*see* Figure 6-2).

Figure 6-2. A Negative and a Positive Image

Negative   Positive

Reversal film, on the other hand, will have the "real" values after developing the film. The negative to positive, reversal process is done chemically. This type of film must be viewed with the aid of a projector. Slides are examples of this film group.

Film Speed

Films have a "speed"; they react to light at different rates. Film speed is known in this country by an ASA number (the ASA stands for the old American Standards Association). In other countries, they may have a DIN number (DIN — Deutsches Industrie Norm). Most films available have both numbers, and care should be given to use the proper guide number to determine the correct exposure. These numbers will be clearly printed on both the film package and the film roll cover (*see* Figure 6-3, page 140). The film EI number (Exposure Index), as it has been termed more recently, will tell us how quickly or how slowly the film reacts to light; it indicates the film's sensitivity to light. A film marked ASA 64 is slower than ASA 125 film; ASA 400 film is faster than ASA 125 film. The higher the number, the faster the film.

Image Sharpness

The speed of the film bears a relationship to its chemical composition — a faster film has coarser chemical components that appear as "grain." Grain will determine the sharpness of the reproduced image. To reproduce greater detail, a slow film is necessary; a fast film will need to be used when the available light is not sufficient to expose slow film well. In this case, detail is sometimes sacrificed to gain speed.

Figure 6-3. 35mm Film Spool and Container

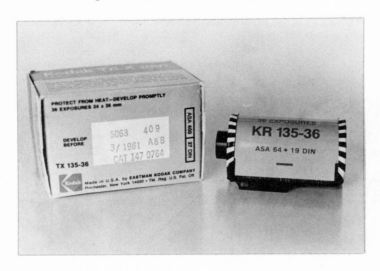

Film Size

Film also comes in different gauges or sizes. It is fair to expect that in your work you will encounter mainly the most popular size, the 35mm format. The most usually found sizes are:

110   smallest size of film, instant-load mini cameras
126   cartridge film used in instant-load cameras
135   most commonly found film in the amateur and professional field, referred as 35mm
2x2   larger size, mainly for expensive box cameras
4x5   strictly for professional use
8x10 strictly for professional use

The following chart lists some readily available 35mm film in 20-exposure or 36-exposure rolls supplied by Eastman Kodak.

Negative Film

Name	Color	Speed ASA	Grain
Plus-x Pan	B and W	125	Extremely fine
Tri-x Pan	B and W	400	Fine
Kodacolor II	Color	100	Microfine
Kodacolor 400	Color	400	Fine

Reversal Film

Name	Color	Speed ASA	Grain
Kodachrome 25	Color	25	Extremely fine
Kodachrome 64	Color	64	Extremely fine
Kodak Ektachrome 64	Color	64	Very fine
Kodak Ektachrome 160	Color	160	Fine
Kodak Ektachrome 200	Color	200	Fine

Summary of Film Exposure

We have encountered four variables that affect the exposure of film:

1) Light intensity
2) Lens aperture
3) Shutter speed
4) Film sensitivity

Different light sources give off light of varied intensity and a light or exposure meter is used to measure the strength of the light. The lens aperture is controlled by a ring marked with *f*-stop numbers. How long the shutter reamins open is timed in fractions of a second. Film is available with different emulsions, and their sensitivity to light is indexed with ASA or EI numbers.

LIGHT AND ITS COLORS

The experiments Issac Newton carried out in 1666 may be well known to you. His prism showed unmistakably that white light is made up of different colors. However, not many of us know that different sources of what appears to the human eye as white light are not recorded as such on the photographic emulsions of films. Film emulsion must be matched to the natural or artificial illumination if we wish our photograph to have a natural color balance. To select film, we must learn to identify the colors present in the available light. To simplify this task, light has been assigned a color temperature scale. This scale is measured in degrees Kelvin (°K) and will indicate the dominant color wave length present in a light source. Although *colorimetry* is a rather complicated subject, awareness of some principles and a few facts will allow us to obtain good color photographs. Figure 6-4, page 142, shows approximate color temperature for some commonly available light sources. These values are not absolutes. The color temperature of photoflood bulbs, for instance, changes with use, and natural light is affected by atmospheric conditions. An open fire may be strong or weak. Nevertheless, one fact is easily observed in looking at the diagram: a low color temperature light source has more red wavelengths and a high color temperature source has more blue wavelengths.

MATCHING FILM TO LIGHT

Exposing black-and-white film presents no difficulties with the color temperature of the illumination source. There *are* some fine chemical

Figure 6-4. Color Temperature of Several Light Sources

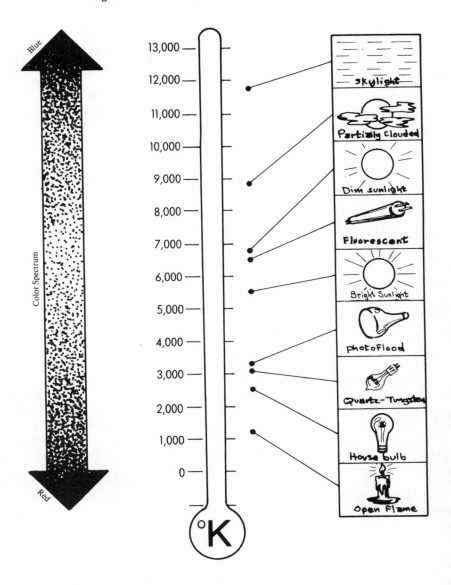

considerations relating the sensitivity of black-and-white emulsions to the light wavelengths, but they are of little importance to us—for our needs, black-and-white photographic film can be exposed with any light source, and the color of the light can be disregarded.

Color film selection is another issue, and we must examine the two types of illumination: artificial and natural. Sunlight will be recorded on daylight film with the correct color balance under normal conditions—we will indicate exceptions later. For indoor photography, we may find mainly five different artificial light sources:

1) Flash
2) Photoflood 3, 400 °K
3) Quartz-tungsten 3,200 °K
4) Fluorescent
5) Incandescent house bulb

The *flash* or *strobe* is a substitute for natural lighting. It has a color temperature similar to that of daylight; therefore, daylight film is used with it. The name "flash" implies a single occurrence of light output. This characteristic makes the predictability of accurate exposure harder to achieve without auxiliary measuring devices. The light intensity of flash guns or strobes cannot be changed, and each available type has a different light intensity. Therefore, the flash unit's distance to the object and the lens aperture are the two variables that will affect exposure. Practice and following the manufacturer's instructions will afford good exposures.

Photoflood bulbs are used in photographic copy stands and photographic studios (*see* Figure 6-5). These lights will be marked 3,400 °K. Although their color temperature changes with use, they keep color temperature constant for a good part of their useful life. They should be discarded or used for black-and-white work either when they are too old or when their use time is unknown. They can also be tested by exposing film using them and examining the resulting color rendition. A color balanced film has been developed for use with these 3,400 = °K photoflood lights. It will be identified as either professional type "A" or simply 3,400 = °K.

Figure 6-5. Typical Photoflood Bulb

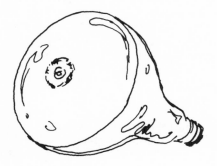

Quartz-tungsten bulbs are also used in copy stands, but usually only in those of the more expensive professional type. They are smaller in size than photoflood bulbs. They may have different shapes and are identified 3,200 °K (*see* Figure 6-6).

Figure 6-6. Quartz Bulbs

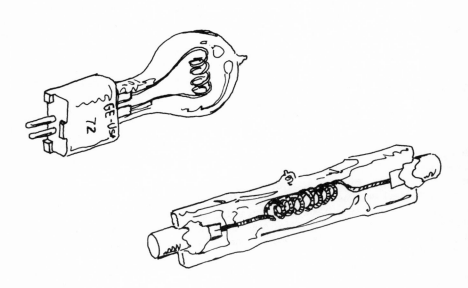

These bulbs are very bright and their color temperature changes only slightly during their life time, which is an advantage when good color balance is always required — the color of their light is predictable. The film to be used under this type of illumination is type "B" or tungsten 3,200 °K.

No film is specifically color balanced for the remaining two forms of artificial illumination: fluorescent and incandescent house bulbs. Fluorescent lights are a nuisance in photographic work. A great variety of them are available, and the color of their light changes with every manufacturer. The color tends to be concentrated around the green-blue portion of the visual spectrum. There are some manufacturers that claim theirs are equal to daylight, but none will affect film as sunlight will! Fluorescent bulbs also change the color response as they heat up and as they age. But, we have another way to handle them that, if not perfect, is adequate. We use a filter.

FILTERS

Photographic filters are either tinted, round pieces of glass or gelatin square that are placed in the light path to "correct" the color of light. By using filters color can be "subtracted" or "added." The filter accepts or transmits the color wavelength that is visible. For example, if we look at a filter and it looks yellow, i will let the yellow component of the light pass through and will reject the blu part of the visual spectrum. Without getting deeper again into highly technica matters, let us go back and "control" the color temperature of fluorescent lights using these filters.

A combination of two filters, a 20B plus a 5M, will correct or balance the light from a fluorescent tube. The number in the filter indicates the "strength," the degree of filtration. The letter stands for the initial of the corresponding hue: B = blue, M = magenta, R = red, etc. The appropriate combination is already found on manufactured filters specially designed for this purpose. An FLD filter should be used if illumination is by *F*lourescent *L*ight and the film is *D*aylight type. An FLB filter should be used when the film is type *B* or balanced for 3,200 °K. In the case of household incandescent bulbs, the filter should be blue. The color quality of light is so variable that it is hard to have a "perfect" filter, and, again, expert handling of filters requires a good amount of experimentation and knowledge. For your purposes, consult the color temperature and filter charts below to find out what film to choose for the available light and what filtration may be necessary. This information may be beneficial because you may have loaded the camera with daylight film, and in the middle of the roll you may be requested to shoot a scene in an underground science lab. Without wasting film, or getting people with green faces, you may compensate and turn out a professional looking job.

Film Type	°K
Kodachrome 25	Daylight
Kodachrome 64	Daylight
Ektachrome 64	Daylight
Ektachrome 200	Daylight
Kodachrome 40, Type A	3,400
Ektachrome 160, Type B	3,200

Color Temperature of Some Kodak Color Films

Film Type	Illumination	Filter
Daylight	3,200 °K	80 A
Daylight	3,400 °K	80 B
Daylight	Fluorescent	FLD
Type A, 3,400 °K	Daylight	85
Type B, 3,200 °K	Daylight	85 B
Type B, 3,200 °K	Fluorescent	FLB

Required Filters

CAUSES OF POOR PICTURES

What happens if you use the wrong film? These are some of the results:

- Daylight film exposed under fluorescent light will have a greenish-blue tint.
- Daylight film under household bulb illumination will have a yellowish tint.
- "B" type film (3,200 °K) exposed to daylight will have a blue tint.

We mentioned earlier that filtration may also be necessary when the right film is used with light that is apparently right. For example, a daylight film used on a sunny day to photograph a snow covered landscape will have a bluish cast to it. Why? Snow is white and highly reflective. It will reflect the skylight from above into the camera. The result is a dominant blue picture from the high color temperature sky. Remember, the sky is around 11,000 °K. Just follow the charts, and when you have time to experiment, have fun! Some of the results can be rewarding and exciting.

LENSES

We have indicated that a camera lens is used to obtain a sharp image. Simple and inexpensive cameras have a fixed lens. This lens has only one piece of glass, or *optical element*. There is no need to focus the lens to obtain a sharp picture. More expensive and professional cameras have interchangeable lenses. These lenses are made with several optical elements, one or more of which must be shifted or moved to focus the image. These lenses have *selective* focus capabilities—the focusing distance can be changed. You may focus on an object 10 feet away or on an object 2 feet away.

Focusing

Focusing is done by rotating a ring usually placed in the front of the lens body. To aid focusing, the camera *viewfinder* has an inner glass screen with other optical elements. The most commonly found screens are the *split-prism* and the *multi-* , or *micro-prism* screens. The split-prism screen will let you see the object on which you focus, "splitted." As you focus the lens, the parts of the object come together on the screen. At the point of maximum sharpness, the object will look "normal" or "put together" (*see* Figure 6-7). On a multi-prism screen (*see* Figure 6-8), the scene will appear to be "vibrating," "moving," if it is out of focus. As the lens is focused, the movement "stops"; the scene is "quiet" at its focus point. Not only will the object on which you focus appear sharp in your photograph, other objects in front of or behind your selected focus point will also appear sharp. This area, or distance, is known as the *depth of field*.

Figure 6-7. Split-prism Screen

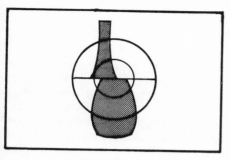
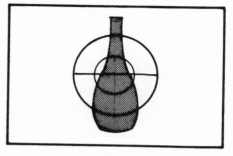

Out of Focus

In Focus

Figure 6-8. Multi-prism Screen

Depth of Field

Depth of field is the space in back and in front of your focused object in which other objects will also photograph in focus. (*see* Figure 6-9).

Figure 6-9. Depth of Field

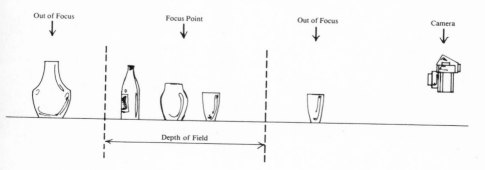

The depth of field is controlled by:

1) subject-to-camera distance
2) lens aperture
3) type of lens

Subject-to-camera distance is an important consideration in determining the depth of field of a photograph. Lenses have the greatest depth of field when focused at infinity. Photographs made out-of-doors, such as scenic views, will be in focus from some point a few feet in front of the camera to the distant horizon. (A roman numeral eight on its side is the sign for infinity printed on most cameras with selective focus capabilities.) On the other hand, the closer the object is to the camera the shallower the depth of field will become. If the focusing point is three feet from the camera, the depth of field may be only a few inches.

Lens aperture is our next concern in determining depth of field. A small lens opening will have a greater depth of field at any focusing distance than will a larger f-stop. For instance, an $f/16$ stop may have a depth of field of three feet when the lens is focused at five feet (*see* Figure 6-10).

Figure 6-10. Depth of Field with Small f-stop

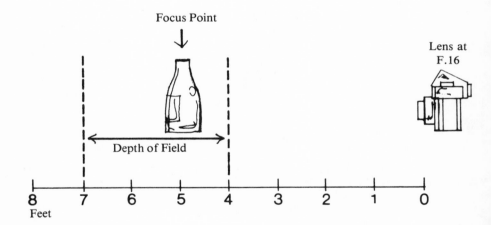

In contrast, the same lens at the same focusing distance may give a depth of field on one foot when opened to $f/2.8$ (*see* Figure 6-11). Remember, that the smaller the opening of the lens, the larger the depth of field. You may use the selective capabilities of your lenses in creating effects for your photographs. By applying these principles you can control your photographic designs. As an example, if you wish to have a person in the foreground in sharp focus with a background out-of-focus, you need to get close to the person (small camera-to-object distance) and use a large f-stop (lens aperture). Combining these two variables with the variable of focal length will give you greater flexibility in establishing the depth of field of your photographs.

Figure 6-11. Depth of Field with Large *f*-stop

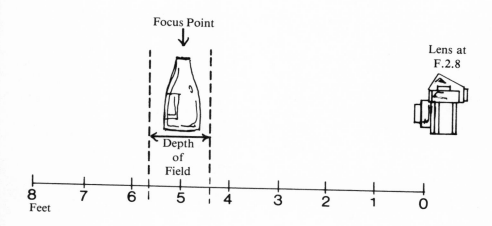

Focal Length

Interchangeable lenses can be obtained in different focal lengths. Simply stated, the focal length of the lens is determined by the distance of some of your lens elements to your film plane or surface. This distance determines the "optical size" of the lens. The focal length of a lens is measured in millimeters; typical lens sizes are 35mm, 55mm, 80mm, 105mm, 200mm, 400mm, 800mm. Lenses 35mm and smaller are said to be "short" lenses. The "normal" lens for 35mm cameras is a 50mm lens. Lenses over 50mm are considered "long" lenses. "Long" lenses give shallower depths of field than do "short" lenses. Looking at our earlier example, a 50mm lens at $f/16$, focused at five feet may give a four-foot depth of field (*see* Figure 6-12).

Figure 6-12. Depth of Field with 50mm Lens

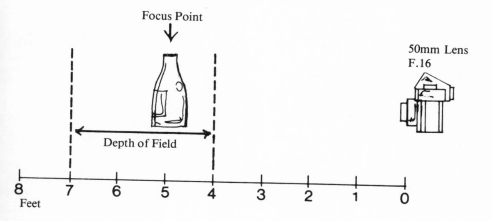

Under the same conditions a 105mm lens may give a depth of field of only a few inches (*see* Figure 6-13).

Figure 6-13. Depth of Field with 105mm Lens

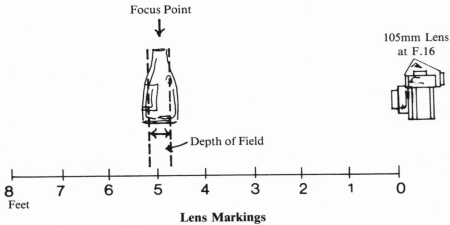

Lens Markings

Variables that control the depth of field are, in most cases, marked on your lens. A typical lens will have on its focusing ring a number scale indicating distance. This scale will be opposite a set of markings that correspond to the *f*-stop numbers of the aperture control ring. Between these two rings, a series of lines will mark the depth of field (*see* Figure 6-14).

Figure 6-14. Lens Markings

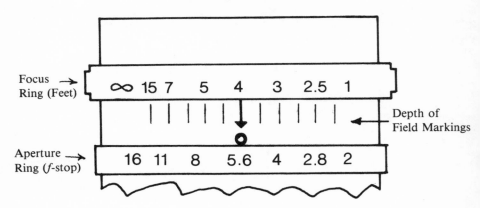

Let's interpret this information (*see* Figure 6-15). In our example, the arrow in the center scale points at *f*-stop 5.6 and at 4 feet: the lens is set at a focusing distance of 4 feet and at *f*/5.6 aperture. (In our diagram the lines are black; on some lenses they are color-coded in relation to the aperture colors.) Let's assume that the first two lines left and right of the arrow indicate *f*/2, the next two *f*/2.8, and so on. Looking at the lines that correspond to *f*/5.6 (the fourths to either side of the arrow) we see them opposite 2.5 feet and 7 feet. This indicates that the depth of field is 4½ feet long and begins at 2.5 feet from the lens and extends to 7 feet from it.

Figure 6-15. Depth of Field Lens Markings

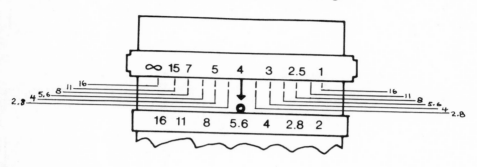

Let's look at another example without changing the lens focusing distance, but opening up the lens to *f/2.8* we will discover that the depth of field will be from approximately 3½ feet to 4½ feet in front of the camera, or a mere one foot (*see* Figure 6-16).

Figure 6-16. Finding the Depth of Field Using Lens Markings

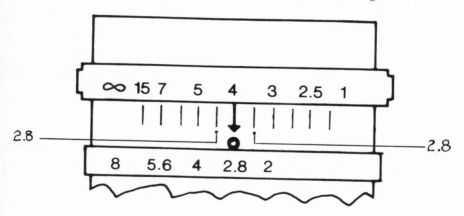

Simple, isn't it? Practice this concept by looking at interchangeable lenses and selectively focusing on objects at different distances while changing the lens openings.

Lens Coverage

The optical size of the lens determines the scene that you can photograph at a given distance. The area or angle of coverage will also be selected by the choice of lens. A "short" lens is a "wide" lens—the angle of view of such a lens is wide. In contrast, a "long" lens has a narrow angle of coverage (*see* Figure 6-17, page 152).

Figure 6-17. Area of Coverage with Lenses of Different Focal Lengths

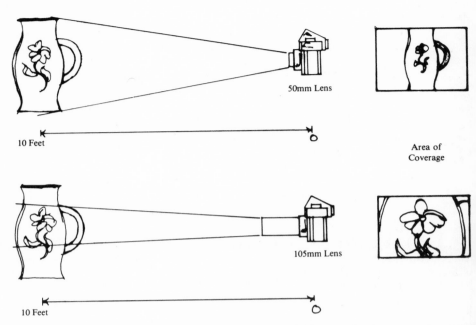

50mm Lens

10 Feet

Area of
Coverage

105mm Lens

10 Feet

Look at the same scene when photographed with different size lenses at the same camera-to-object distance (*see* Figure 6-18). Obviously, lens optical size will be chosen considering the distances you may have available. For example, if you would like to get a close up photograph of a wild animal, you will need a long lens if only because the animal won't let you come near. If you have a wide lens, the resulting shot will be a panoramic view, and the animal will usually appear as a spot.

Variable Focal Length Lens

A lens that can change the area of coverage is called a variable focal length lens or *zoom* lens. Again, by rotating a ring on the lens body certain optical elements will change their place in the lens and allow you to select the angle of view. Zoom lenses are identified by two sets of numbers, e.g., 50-100mm. These numbers indicate the extremes of the range. In this case, the lens' widest coverage will be that of a 50mm lens, and that of a 100mm lens will be its narrowest angle.

PHOTOGRAPHING HINTS

Now that we have reviewed several facts about photography and its principles, we should put our knowledge to use.

- Familiarize yourself with the camera you are to use for an assignment ahead of time. There are many different makes and types of cameras. The degree of sophistication, and at times complexity, present technology has reached provides many

Figure 6-18. Comparing Coverage of Several Focal Length Lenses

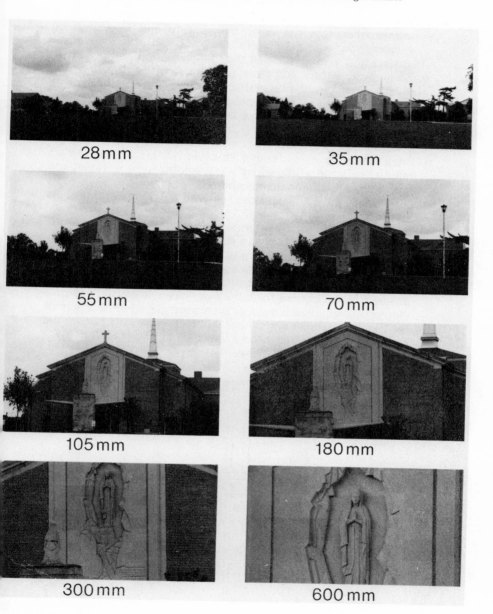

different models. Make sure you can identify all parts and controls before loading the film in the camera.

- Select your film and load it. Remember to select the film based on available light (its color temperature and intensity), on whether you will need color or black-and-white, slides or photographs. Make sure that the film is inserted on the take-up mechanism. Before you close the camera check that the film advances well on the film path and that *sprockets* in the take-up spool fit into the sprocket holes.
- Set your exposure meter according to the EI or ASA film index numbers.
- Select your lens or lenses. If you wil be shooting away from your work station, bring additional lenses if available. Unless you are familiar with the setting where you are going to be working, you may find a need for an unexpected space coverage. In most cases, the work is done in closed environments, and a wide angle lens is helpful.
- Bring a tripod along if you suspect the available light to be insufficient. Remember that a shutter speed of 1/30 of a second or slower requires a steady camera to eliminate blurred pictures.
- To duplicate flat materials such as book pages, charts, drawings, etc., a copy stand is used. Such a stand allows you to adjust the distance of the camera from the *bed* or copying surface to provide sufficient space coverage. Lights are placed at each side of the flat materials for even illumination. To eliminate glare in the case of glossy materials, these lights should be placed at a 45° angle to the bed (*see* Figure 6-19).

Figure 6-19. Lights on a Copy Stand

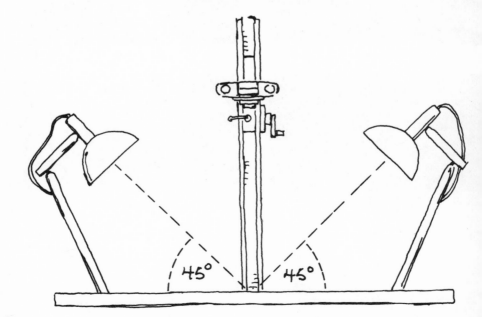

A tripod can duplicate the services of a copy stand. Materials can be placed on a low table or even on the floor. A pair of portable lights can be used for illumination.

- In copying materials, a background is necessary which will not interfere with the object. Cover the bed adequately to provide the object with a uniform, unobtrusive backing.
- Exposed film should not be stored for long. It should be processed as soon as possible. Exposed film will change its chemical composition quickly (especially color film). If the film cannot be processed after exposure, it should be refrigerated to slow down chemical reaction. Be careful to place the film in an air-tight container, such as a plastic bag, before placing it in a refrigerator. This precaution will eliminate condensation on the film, which may appear as water spots on the photographs.

MULTIMEDIA INSTRUCTIONAL MATERIALS

Photography. McGraw Hill Films, 1968. A series or single concept 8mm loops, silent, B&W.

Write the Eastman Kodak Company, Audio Visual Services, Rochester, NY. It maintains a loan collection of films and slide/sound presentations describing the technical aspects of photography.

BASIC SOURCES

Craven, George M. *Object and Image: An Introduction to Photography.* Englewood Cliffs, NJ: Prentice-Hall, 1975.

Encyclopedia of Practical Photography, 14 vols. New York: Kodak Corp., 1978.

Life Library of Photography. New York: Time and Life Books, 1977- .

First Phonograph Invented by Thomas A. Edison

The original tin-foil phonograph invented by Thomas A. Edison in the autumn of 1877.

(This photograph has been made available through the Thomas Alva Edison Foundation by the Edison National Historic Site)

7 – AUDIORECORDINGS

Man's innate need to communicate led first to the spoken word, then to graphic symbols, which allowed speech to be preserved. Writings and drawings have left a visual record of past eras, and printing has made that record available to the public. But, compared to the vitality of speech, writing is often static and the record is often inaccurate. It was late in the nineteenth century when sound were for the first time dynamically and faithfully recorded. In 1877, Thomas A Edison created his phonograph (*see* Frontispiece), a device that made possible the recording and reproduction of the human voice. This dynamic invention recorded sound by inscribing with a vibrating stylus a very narrow groove on a thin sheet of tin foil wrapped around a cylinder. The original sound could be reproduced by having the needle retrace its path over the groove. This invention led to other and better systems. Emile Berliner used a stylus that scratched a wavy line on a flat zinc disc covered with a greasy substance. Etched in acid, this grooved disc was used as a master to cast a facsimile on a hard, rubber-like material. This system eventually gave us today's phonograph records. To these mechanical devices other systems were added that employed electromagnetic principles. First came magnetic recordings on thin wire, later magnetic substances on paper tapes, and now plastic magnetic tape used for sound *and* for visual recordings.

SOUND

Sound is the result of variations in air pressure caused by a vibrating object. Any object moving back and forth pushes the air around it generating air waves that will be sensed by the human ear as sound.

Sound Waves and Frequency

Using a simple string instrument, a rubber band stretched between two fingers, we can graphically plot a sound wave or cycle (*see* Figure 7-1). Observing the vibration of the rubber band after plucking on it, we see that the band rises to a high point, and then bounces back to a low point always passing through the rest position. The cycle can go on and on.

Figure 7-1. Simple Example for Plotting a Sound Wave or Cycle

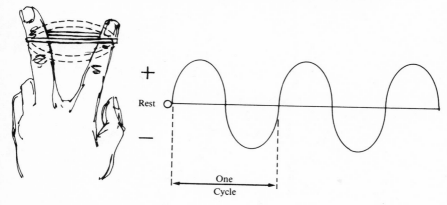

Counting the number of cycles per second would give the repetition rate or *frequency* of the sound. A slowly vibrating object will produce sounds of low frequency, or in the bass range of the audio spectrum. In contrast, a rapidly vibrating object will produce sounds of high frequency, or in the treble range. The human ear is capable of detecting a wide range of frequencies. The average ear hears from about 20 to 20,000 cycles per second. Cycles per seconds are called Hertz (e.g., 10 cycles per second = 10 Hz.).

Sounds are not usually made up of one frequency. A musical instrument generates a wide variety of frequencies, and these frequencies alternate very rapidly. The human ear responds very quickly and is able to distinguish all these variations. When sounds are recorded and reproduced by electro-mechanical equipment, this equipment must be able to render all frequencies in the original sounds. *Fidelity* refers to the accurate reproduction of all the frequencies present in the original sound.

Microphones

A microphone is the electronic equivalent of the human ear. A microphone will detect variations of air pressure and transmit this information to the "brain" of the sound system in the form of an electric current. Microphones are *transducers* of energy; they transform mechanical energy into electro-magnetic energy.

Microphone Types

A basic experiment will show that when a magnet moves back and forth next to a coil of wire, an electric current flows through the wire. A *dynamic microphone* uses just this principle (*see* Figure 7-2). A membrane or diaphragm vibrates in response to changes in air pressure. This diaphragm has a voice coil attached to it, and it is enclosed in a magnetic structure. As the voice coil moves back and forth with the diaphragm, it generates an electric current that will be transformed into a voltage directly related to the original sound wave—i.e., a loud sound will generate a high voltage and a weaker sound will produce a lower voltage output.

Figure 7-2. Basic Dynamic Microphone

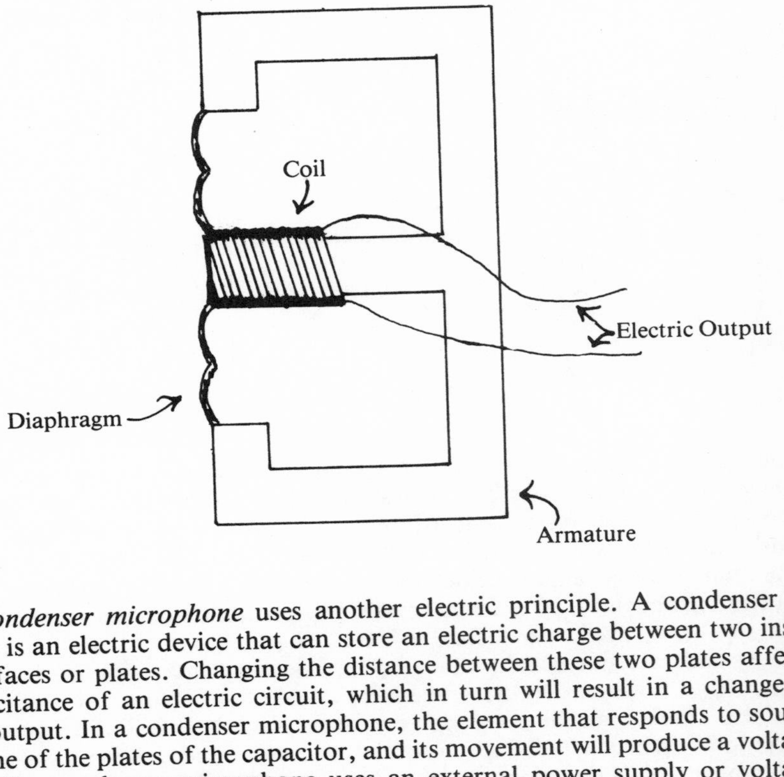

A *condenser microphone* uses another electric principle. A condenser capacitor is an electric device that can store an electric charge between two insulated surfaces or plates. Changing the distance between these two plates affects the capacitance of an electric circuit, which in turn will result in a change voltage output. In a condenser microphone, the element that responds to sound acts as one of the plates of the capacitor, and its movement will produce a voltage change. The condenser microphone uses an external power supply or voltage source. A new type of condenser microphone, the *electrect* uses an internal battery in most cases.

In addition to having to respond to a wide range of frequencies microphones have to be sensitive to and selective of sounds. Sound travels in directions, bounces off most surfaces, and passes through solid bodies. Sound can also be very loud or very soft. Microphones reject or accept sound frequencies in accordance with their physical and electronic design. Some will be more responsive to a portion of the audio spectrum while others will discriminate between sounds on the same audio range. The way in which microphones "select" "discriminate" sounds brings us to another characteristic of microphones: their *pickup* patterns.

Pickup Patterns

The sensitivity and selectivity of a microphone is plotted on *polar* patterns *(see* Figure 7-3). The understanding of these patterns will help us select the right microphone for our needs.

Figure 7-3. Polar Pattern of Microphone Pickup

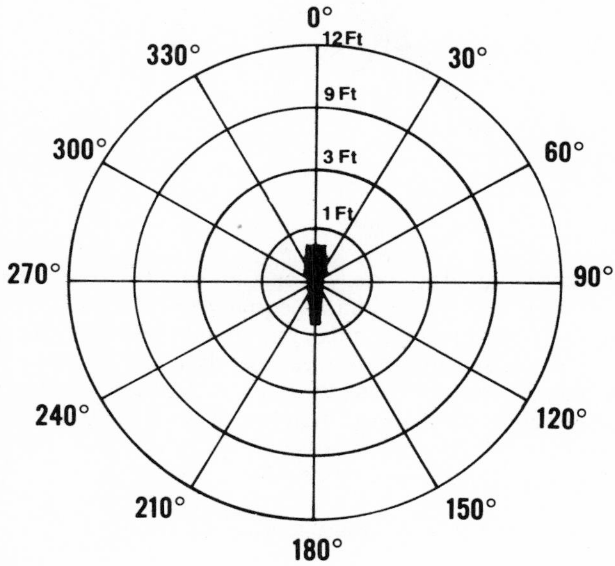

The front of the microphone indicates the 0 axis. The concentric circles indicate the sensitivity in decibels (db), which are the units of measurement for sound. 0 decibel corresponds not to absolute silence, but to the human threshold of hearing. The distance from one circle to another is usually marked in feet. The microphone polar pattern will show lines that correspond to the response at different frequencies. These alternations are indicated in hertz (Hz — single cycle), kilohertz (Khz — one thousand cycles), or megahertz (Mhz — one million cycles). Reference to megahertz will be encountered in higher frequency equipment for purposes other than audio-engineeering. The samples below are for illustration only, and do not represent any microphone in particular.

These pickup patterns seem to show that microphone sensitivity occurs on a flat area in front and around it, but these patterns are to be interpreted in three dimensions *(see* Figure 7-4, page 160). Learn to visualize this tridimensionality when you interpret the polar pattern in selecting the microphone suitable to your needs. Microphones come in three basic types of pickup patterns: *omnidirectional, unidirectional* or *cardioid*, and *bidirectional*.

Figure 7-4. Three-dimensional Nature of Microphone Sensitivity

Omnidirectional

The omnidirectional microphone will pick up sounds coming from all directions — the front, sides, and rear — almost equally well (*see* Figure 7-5). Placed between two persons, for instance, this type of microphone will pick up both voices with nearly equal sensitivity. We say nearly equal because even the best omnidirectional microphone will tend to become directional at certain frequencies; it will discriminate against those frequencies when they originate from the back of the microphone. In an extreme case the quality of the voice coming from the rear may change and will sound "duller."

Figure 7-5. Omnidirectional Pattern

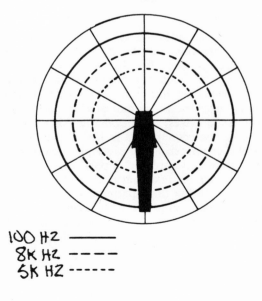

Unidirectional or Cardioid

Unidirectional microphones respond best at sounds coming from the front of the microphone and are far less sensitive to sounds coming from the back (*see* Figure 7-6).

Figure 7-6. Cardioid Pattern

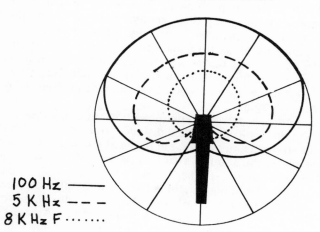

100 Hz ———
5 K Hz — — —
8 K Hz F · · · · · · ·

The resulting shape of the pickup pattern is heart shaped, which is the reason these microphones are termed "cardioid." Just as with the omnidirectional microphones, the cardioids respond better at some frequencies than at others. Only the best and more expensive ones are equally selective over the full range of the audible spectrum.

Bidirectional

Two cardioid microphones placed back to back would have the same sensitivity as a bidirectional microphone (*see* Figure 7-7). This dual receptivity best receives sounds coming at the microphone from the front and back and discriminates against sound coming from the sides.

Figure 7-7. Bidirectional Pattern

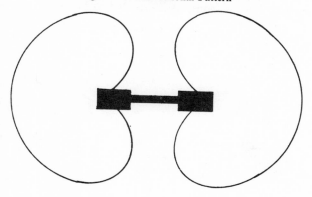

Hints on Microphone Handling and Placement

Microphone placement is an important consideration in sound recording or amplification. Microphone handling is equally as important. First you must learn to handle microphones with care; their diaphragms and armatures are very fragile and sensitive. Microphones should not be dropped or hit. Also, many microphones have elements that will be damaged if exposed to high air pressures. Microphones should not be tested by blowing into them. To test them, it is preferable to speak normally, or to tap them gently. When recording outdoors the use of windscreens is recommended, especially during windy days. This precaution prevents damaging a sensitive microphone and will also eliminate wind sounds. In using an electrect or battery-powered microphone, make sure that the battery is in good condition. This microphone usually has a switch to turn the battery on. After using the microphone remember to switch the battery off to prevent it from discharging.

To record more than one person, omnidirectional microphones are preferred. Microphones should be placed at equal distance from every person if they have voices of the same strength and quality, but if one of the persons has a softer voice, the microphone placement should favor that person. Microphones should be placed at least 12 inches from the speaker's mouth. Close "miking" is only for performers who are used to handling microphones. A microphone placed too close to the person's mouth picks up "pops" and sibilant sounds. These sounds, which accompany initial ps or ss, are emphasized at close range. At short range the person should not talk into the microphone, but across it. This minimizes "breathiness" and s sounds.

Cardioid microphones are preferred in situations where a directional pickup is needed. In noisy surroundings, this quality will discriminate against undesirable sounds. But remember that sound "bounces off" surfaces, therefore even sounds coming from the back of a very directional microphone may find their way to the front in closed spaces.

A situation often encountered in a classroom or a conference is the amplification of sound, which produces "feedback." Feedback is a squealing sound produced when the amplified sound comes out of the speakers and goes back into the microphone again and again, creating a self-reinforcing sound loop (*see* Figure 7-8). It is true that by reducing the amplification volume you will decrease feedback, but this may be at a point where amplification is lower than necessary. In such cases, it is more desirable to use a cardioid microphone to help minimize sound coming at it from the back and to place the microphone behind the speakers, rather than in front, as illustrated in Figure 7-9.

Figure 7-8. Feedback Loop

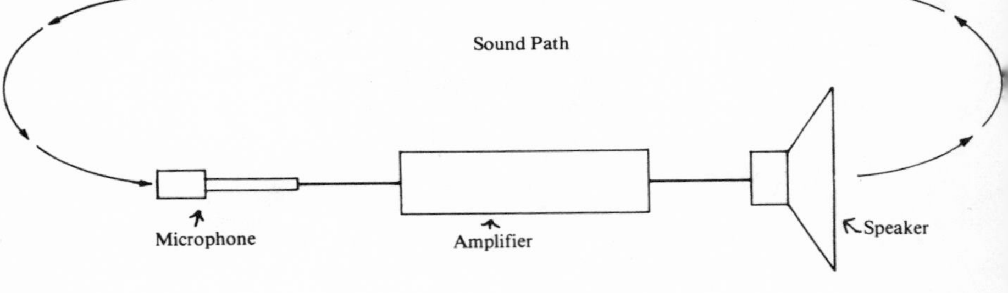

Figure 7-9. Microphone Placement in Relation to Loudspeakers

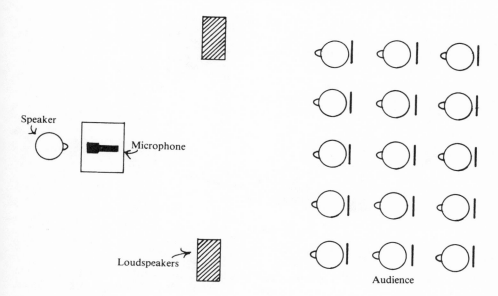

Electronic Characteristics of Microphones

In situations in which more than one microphone is needed, all should be of the same type. This will eliminate problems that may be caused by differences in frequency response, sound quality, sensitivity, and other characteristics such as 1) "impedance" and 2) "balanced" or "unbalanced line" transmission.

Impedance is the resistance that an electronic circuit presents to an electric current. For our purposes we must be aware only that inputs and outputs of electronic gear must match impedances to eliminate distortion in the signal. Microphones are of either low or high impedance. They will be initialied *hi-Z* or *low-Z*. Z is the electronic symbol for impedance and is measured in ohms. The ohm is the measuring unit of resitance and is indicated by the Greek letter omega (Ω). A sample of a typical microphone description will read: low-Z, 600 Ω. Typical low impedances range up to 10,000Ω and high impedances up to 50,000 Ω.

It is important to know that high impedance microphones must be used with the length of cable the manufacturer has provided. This length should not be increased. Additional cabling will introduce distortion and loss of sensitivity. In short, it will affect the efficiency of the microphone and thus the sound quality. In contrast, a low impedance microphone cable can be lengthened without appreciable loss in sound quality.

The terms "balanced" or "unbalanced" transmission line refer to the wiring configuration of the conductor cable. An unbalanced line has two conductors or wires. One conductor is used for carrying the signal, and the second serves as the "shield" or "ground" (*see* Figure 7-10A, page 164). The signal wire may be called the "live" wire. A balanced line has two conductors for the signal transmission and a third shield or ground conductor (*see* Figure 7-10B, page 164).

Figure 7-10. Audio Cables

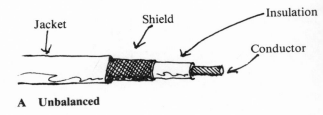

A   Unbalanced

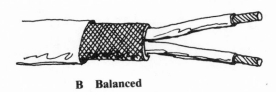

B   Balanced

Most low impedance microphones have a balanced line output. A high impedance microphone usually has an unbalanced line output. One advantage of a balanced line is that it does not pick up hum noises from electric power lines that may be close by. Microphone connectors will give a good clue to the type of impedance and transmission line type. The hi-Z unbalanced microphone usually has a single prong connector; the connector for low-Z balanced line microphone usually has three prongs (*see* Figure 7-11).

Figure 7-11. Most Commonly Available Microphone Connectors

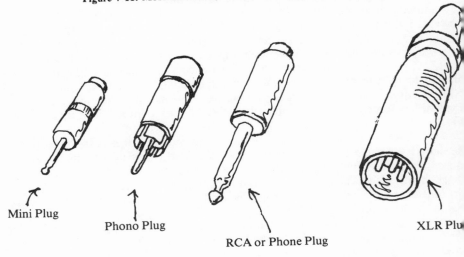

Mixers, Preamplifiers, and Amplifiers

When more than one microphone is required simultaneously, there is the need to combine or "mix" the output signal of all microphones into one composite signal. An electronic mixer is used for this purpose (*see* Figure 7-12). Mixers have several microphone "inputs" with individual volume controls and a master volume control. These individual controls make some signals louder and others softer, which permits balancing the incoming sounds from each microphone. The master volume controls the output signal after it is mixed.

Figure 7-12. Mixer

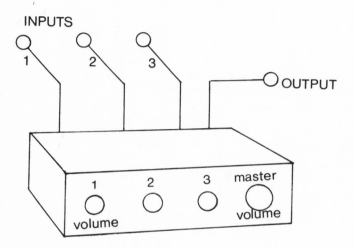

A mixer may have other inputs to accept a signal from a record player, a tape recorder, or a radio. This feature allows the music or sound effects to be added to voices coming from microphone inputs.

Mixers also have tone controls. We have learned that sounds are made up of different frequencies. Electronic equipment such as microphones, mixers, or recorders will emphasize some frequencies more than others. There is no "perfect" equipment that has a flat frequency response. A flat response would mean that all harmonic components of the sound signal would be reproduced and amplified at the same level, but this is not the case. A typical amplification response curve will show "peaks" and "valleys" (*see* Figure 7-13, page 166). In the example provided by Figure 7-13, response is high at certain frequencies and low at others.

In addition, the acoustics of a classroom or a conference hall will affect the character or "tone" of the sound. For example, heavy curtains and carpets will "absorb" high frequencies more fully than low frequencies. Hard or metallic surfaces will react differently. It is desirable to have the electronic capacity to control these variables in sound reproducing conditions. Tone controls afford us this possibility. With tone controls on a mixer or amplifier, frequencies at the high and low end of the audible spectrum can be balanced. Bass controls adjust low

Figure 7-13. Audio Response Curve

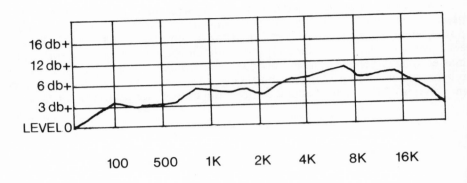

frequencies, and treble controls adjust high frequencies. The equipment can be adjusted to reproduce sound in accordance with our requirements or preferences.

Mixers are also preamplifiers. Microphone signals generally need to be amplified in order to be effectively heard or recorded. These signals are amplified in stages, a little at a time. The amplification level depends on the use of the signal. Signals to be recorded need only preamplification. In this case, the level output of a preamplifier is sufficient to feed a recorder. Further amplification is done at the recorder. Signals to be heard over a loudspeaker need more amplification. These signals are fed from a preamplifier into an amplifier. Amplifiers will strengthen electronic signals to the level necessary to drive speakers. This system of sound processing is usually called a public address system, or P.A., for short. A small P.A. system for a room with 20 people requires less amplification than does one for a large conference hall with 500 persons in it. Amplifier strength is rated in watts, the unit measurement for power. A 5- or 10-watt amplifier is usually adequate for the first situation. In the second case, a 35- or 50-watt unit will be required.

A mixer output signal is called a "line output" when it requires further amplification, and a "speaker out" when it is strong enough to be connected directly to a speaker. A line out can be either a low impedance (600 Ω) balanced line or an unbalanced high impedance output; a speaker out is usually a high level, 8 Ω line. A mixer may also have an "earphone out" to permit a technician to monitor the sound signal.

Again, remember the importance of proper impedance matching in electronic gear to prevent distortion. Hi-Z outputs must be fed to hi-Z inputs. Low-Z outputs may be fed to hi-Z inputs if the mismatch is less than 10 times (i.e., 600 Ω into 1000 Ω), but the contrary is not possible — hi-Z out cannot be fed into low-Z in. In addition to impedance matching, loudspeakers must match the power ratings or the amplifier. The power handling ability of the speakers has to be greater than the power output of the amplifier.

It is necessary to point out here that we have referred to different functions of audio processing equipment that may be found in a single unit. That is to say, preamplifiers, mixers, amplifiers, and even speakers may be enclosed within a single box, or each function may be carried out by a physically independent unit.

Summary of Sound

So far, we know that:

1) Sound is movement of air produced by objects vibrating at different frequencies.

2) Microphones have different pickup patterns and reproduce sounds at a very low signal level.

3) Mixers combine different electronic sound signals and preamplify them to useful processing levels.

4) Amplifiers strengthen preamplified signals and make them audible by driving loudspeakers.

5) Signal lines must match impedance and power to reproduce undistorted sound.

MAGNETIC RECORDING

Magnetic recording and reproducing is accomplished by employing the properties of magnetism. Magnetism polarizes ferrous substances. A magnet can be natural or created by an electric current. We introduced the dynamic microphone by recalling that a magnet moving close to a coil of wire produces an electric current. Conversely, an electric current flowing through the wire coil will create a magnetic field around it. In a magnetic recorder, an electric current from a microphone is amplified and fed to a recording head. This *head* is a metallic armature with a gap and a coil of wire around it (*see* Figure 7-14).

Figure 7-14. Audiorecording Head

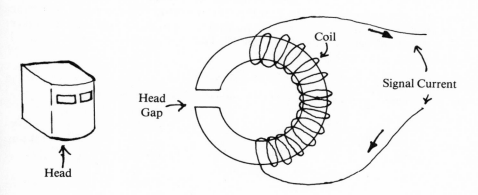

When current flows through the coil, a magnetic field is created across the gap. A magnetic recording tape, a plastic ribbon coated with ferrous oxide particles, passing through this magnetic field will react to it. The metallic particles will be magnetized or "polarized," in accordance with the changing magnetic field produced by the alternating electric current flowing through the head.

Thus the recording tape will receive a magnetic imprint of the signal source. The magnetic tape will preserve this electronic recording indefinitely if the tape is not exposed to another magnetic force. In the playback mode the reverse occurs.

The microscopic magnetized particles on the tape moving past the head gap induce an electric current on the wire coil directly proportional to the recorded information on the tape. This current will be amplified and fed to a loudspeaker, the electronic equivalent to the human vocal chords and mouth. Although based on basic magnetic principles, recording technology requires sophisticated design of functions and components for the production of recording equipment. Besides electronic functions, magnetic recorders have mechanical functions that affect recording quality.

Tape Speed

An important function a recorder must perform well is tape transport. The magnetic tape must pass the recording head at a constant speed to achieve good recording and playback. The speed of the tape also determines recording fidelity. The higher the tape speed, the better the frequency response of the recording. Recording speed is measured by indicating how much tape passes the recording head in a second. The common speeds in audio recorders are 1⅞, 3¾, 7½, and 15 inches per second (ips). Since a faster tape speed results in more tape needed per recordings, the highest speeds are used only in musical and professional recordings. The dynamic range of frequencies in music is greater than in human speech. Therefore, speeds of 15 ips and 7½ ips are used for high fidelity recording: 1⅞ ips and 3¾ ips are used for speech recording.

Format and Size

Recording tape is made with a ribbon of acetate or polyester with magnetic particles bonded to one of the surfaces. The side with the oxide coating looks duller than the backing material, which is usually glossy. The most commonly found audiotape format for audiorecordings was, and still is, the ¼-inch wide tape. This tape comes in open reels and is used in reel-to-reel recorders. This type of recorder accepts a reel of tape, a "supply reel," on its left spindle and an empty reel, a "take-up reel," on its right spindle. The tape passes the recording head from left to right (*see* Figure 7-15).

Figure 7-15. Reel-to-Reel Recorder

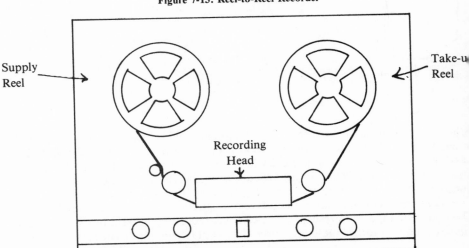

Supply Reel

Take-up Reel

Recording Head

Most recorders accept a 7-inch diameter reel. A few smaller and portable recorders accept a 5-inch reel. Professional studio recording equipment will accept 10¼-inch and 15-inch reels. The diameter of the reel and the thickness of the tape will determine how much tape fits in the reel, and in turn will determine how long a recording can last without interruptions. Thicknesses of recording tape are usually measured in *mils* or thousandths of an inch: .5 mils, 1 mil, 1.5 mil. Thicker tape is easier to handle, is less susceptible to breakage, and stores better. It should be preferred over thinner tape, which should be used only when a long, uninterrupted recording is needed.

The following chart indicates the running time of tapes for the most frequently used tape speeds. In this case the running time indicated is recording or playback of the tape in one direction. The two most common reel diameter sizes are also shown with two different tape thicknesses and capacities.

Reel Dia. (inches)	Tape Thickness (mils)	Tape Length (feet)	Tape Speed (inches/seconds)	Running Time (seconds)
5	1.5	600	1⅞	60
5	1.5	600	3¾	30
5	1.5	600	7½	15
7	1.5	1,200	1⅞	120
7	1.5	1,200	3¾	60
7	1.5	1,200	7½	30
5	.5	1,200	1⅞	120
5	.5	1,200	3¾	60
5	.5	1,200	7½	30
7	.5	2,400	1⅞	240
7	.5	2,400	3¾	120
7	.5	2,400	7½	60

Advanced technological developments in recent years have made possible improvements in recording quality achieved at low speeds and have introduced new tape formats and sizes. Cassette and cartridge tape recorders and playback units are now commonly available. The cartridge recorder is mainly a home entertainment product seldom encountered in other applications. On the other hand, the cassette has found wide application in professional fields due to its ease of handling, portability, and low cost. The cassette has two small reels enclosed in a plastic box with openings on one side that allow a 1/8-inch wide tape to come in contact with the recording head (*see* Figure 7-16, page 170).

The standard cassette is the type "C," which comes in lengths of 10, 20, 30, 60, 90, and 120 minutes of running time at a speed of 1⅞ ips. The cassette shell may be sealed at its seams or have tiny screws to hold it together. The latter is preferred in the event of a malfunction or breakage of the tape, because it can easily be opened for repair.

Another important characteristic influencing recording quality is the tape surface used for recording. The surface area used depends on the recording head size in relation to tape width. There are three types of head sizes usually referred

Figure 7-16. Cassette Tape

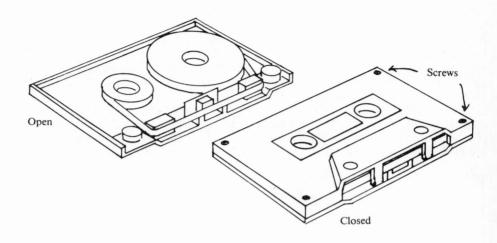

Open

Screws

Closed

to as ¼ track, ½ track or full track heads. The track is the space of the tape width used for recording the signal. With a full track head the information occupies the entire width of the tape (*see* Figure 7-17A).

A ½ track head uses only half of the tape width (*see* Figure 7-17B). It is possible to record on one half of the tape surface and, by flipping the tape over, to record on the other half, in the opposite direction. This can be done as follows: The tape is loaded as usual on the left spindle and run to the end. The tape would then be on the right spindle ready to rewind for playback. Instead, the tape reel is removed from the right spindle, flipped over and run once more. This results in two tracks going in opposite directions. Half track recording utilizes the same length of tape for twice the recording time.

It follows that a ¼ track head will use one-fourth of the tape width (*see* Figure 7-17C). In this case after the first flip of the tape and the second quarter of the track is recorded, it is necessary to reposition the recording head a quarter track down to utilize the center portion of the tape. (*see* Figure 7-18). The same tape length will provide four times as much recording, an appreciable saving of tape. Unfortunately, the reduction in recording surface area will result in a loss of fidelity. This loss has been minimized by the creation of better tape coatings (physical makeup of the tape), better heads, and improved electronics. It is now possible to obtain high fidelity recording and reproduction with ¼ track equipment. The same tape configurations that exist for ¼-inch tape are also found in cassette equipment.

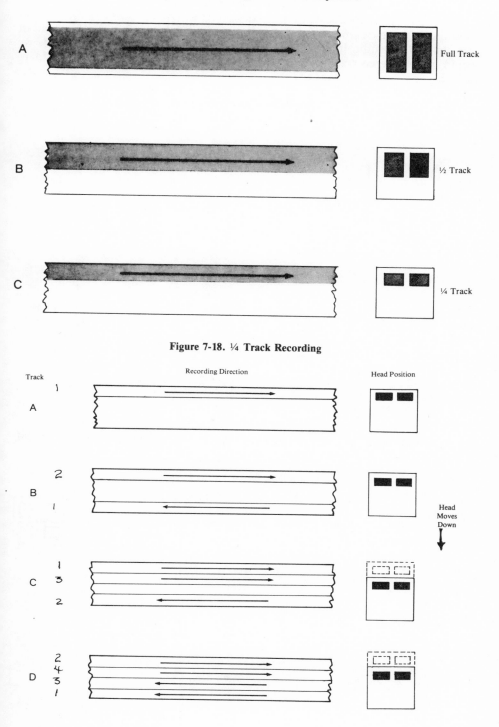

Figure 7-17. Recording Head Size Comparison

Figure 7-18. ¼ Track Recording

Monophonic and Stereo Recording

In the discussions above we have described a signal processing system that employs only one channel of information, or a monophonic system. If instead of one channel of audio information, there were two simultaneous channels, we would have a stereo system (*see* Figure 7-19). Stereo channels are usually identified as right and left channels.

Figure 7-19. Stereo System

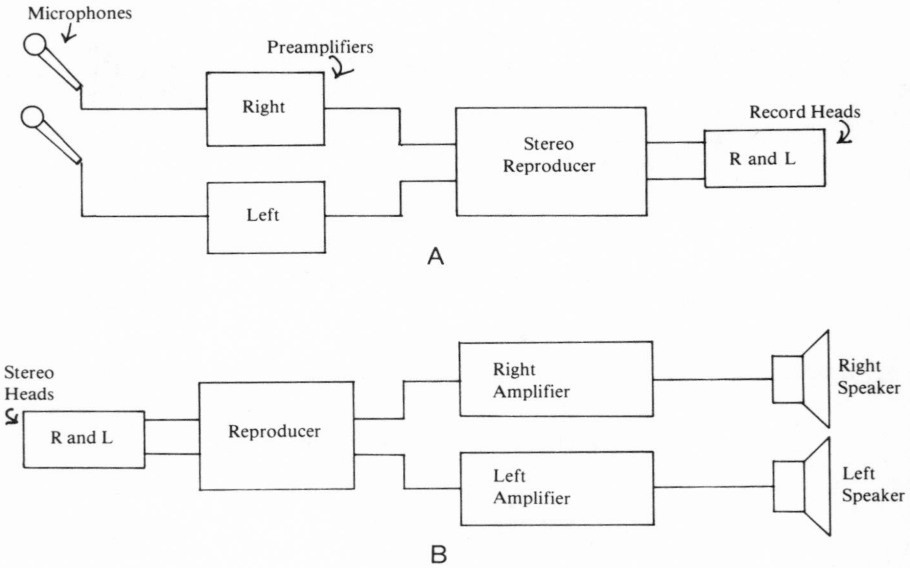

A

B

The ½ track and ¼ track recording head arrangement made stereo recording and reproduction possible (*see* Figure 7-20).

Figure 7-20. Stereo Tracks Layout

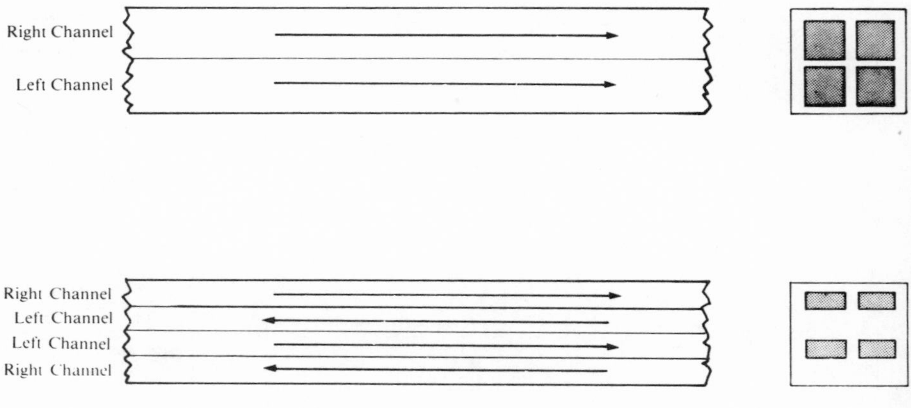

Recorder Functions

In general, recorders have a variety of function controls which vary from manufacturer to manufacturer. Reel-to-reel recorders have switches to control electronic functions combined with levers to control mechanical functions. Cassette recorders have mostly switches controlling both functions. Nevertheless the same basic controls and functions are to be found in all types of monophonic or stereo recorders.

The "forward" or "play" function will activate the transport of the tape for normal play. "Rewind" and "fast forward" will advance the tape at high speeds without reproducing the sound recorded on it. Recording is usually done by activating two switches or levers simultaneously, usually "record" and "forward" or "play." This combined action is designed to eliminate accidental erasures that could be a result of activating only a single function control. "Stop" will halt the transport of tape in the play mode, and in cassette recorders it will also stop the fast forward function. In reel-to-reel recorders the "fast forward" and "rewind" functions are controlled by a lever that must be returned to the neutral position. Many recorders also have a "pause" position. This is different from "stop." In "pause" the tape stops suddenly, but the transport mechanism is not disengaged, as is the case with "stop." "Pause" permits a much smoother stop action.

Most recorders have meters to measure the level of the input signals. The level controls are adjusted for needle fluctuations that reach the maximum recording level mark. Only occasional, brief sound peaks should advance beyond this level. Excessive recording levels will overload the electronic processing components resulting in distorted recordings.

Care of Recorders and Tape

As with any unit performing precise functions, recorders require preventive maintenance and periodic cleaning. The ferrous coating of magnetic tape flakes off as it rubs against metal surfaces. After a few hours of use, accumulations of red or dark brown oxide are usually visible on parts that come in contact with tape. This "dirt" in the tape path, if permitted to accumulate, will result in wrong or erratic tape speed. All surfaces in the tape path should be cleaned by using a cotton swab wetted with isopropyl alcohol of 80% strength. Commercially available magnetic head cleaners of good quality can also be used on all metal parts. They have the added advantage of leaving a lubricant residue to help tape transport and minimize head and tape wear.

Recorders have a rubber pinch roller that keeps the tape pressed against the capstan, a metal revolving rod (*see* Figure 7-21).

Figure 7-21. Audio Recorder Tape Path

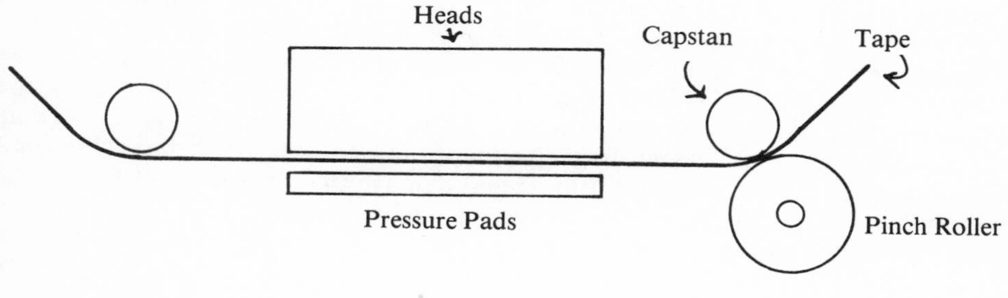

While alcohol will not damage this roller, some head cleaners may. Follow the recorder or the head cleaner manufacturer's instructions in cleaning non-metallic parts. In cleaning, pay special attention to leaving a clean capstan and roller. The capstan's function is to keep the tape speed constant to ensure a good recording. A dirty capstan will contribute to "wow and flutter," distortions resulting from bad tape transport. The supply and take-up spindlers and the capstan, are connected to the electric transport motor by drive belts, or pressure wheels. If these parts are easily accessible they should also be cleaned periodically to ensure good traction. An often-overlooked maintenance step that affects a tape recorder's frequency response seriously is "degaussing" the heads and other metallic parts. Degaussing or demagnetizing metallic parts that come in contact with the tape is essential. Magnetic tape, in passing by metal parts, will polarize or magnetize them gradually. The envelope of the recording head will also be affected by the constant magnetic fields present in its proximity. Magnetism produced in those components must be neutralized. For this purpose a degaussing tool is employed (*see* Figure 7-22).

Figure 7-22. Degausser

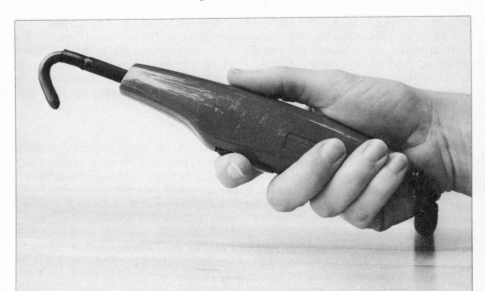

The tool is plugged into a power outlet and turned on away from the recorder. It is then moved toward the recorder, and its tip is brought into close proximity with the head and other metallic parts in the tape path. *The tool must always be in motion* and never stationary during the degaussing operation. After several passes, the tool is taken away from the recorder, at least 4 feet away before it is switched off. This prevents the collapse of the magnetic field at the turning off point to leave a stronger magnetic "imprint" on the recorder's metallic components. Be careful, also, not to heat and damage the recording heads with the metallic tip of the degaussing tool. Many tools have a rubber-covered tip to prevent this from happening.

Magnetic tape is fragile and requires proper handling. It should not be stored in hot or humid environments. The bonding that holds the ferrous oxide to the backing and the backing materials themselves are affected by extreme atmospheric conditions. Also, do not place recorded tape close to strong magnetic field sources such as transformers, motors, etc., which may erase the information on the tape. Wind the tape well and not exceedingly tightly. A recorder with a good transport mechanism winds tape adequately. Take care not to spill tape on the floor, and prevent dirt from accumulating on the tape surface. Dirt is abrasive and will deteriorate the tape coating and wear off recording heads unnecessarily. To mend tape, use splicing tape; do not use other bonding agents that are not designed for this purpose.

Tape Splicing

Recording tape may need splicing, joining two ends together, when the tape is cut due to rough handling or for editing purposes. Although the tape is strong enough for normal machine transport handling, sudden stops of the reels, spool spillage, etc., could cause the tape to break. It may also be desirable at times to condense a speech to only the most important sentences; to arrange musical pieces in a different sequence; or simply to bring together one or more selections that were recorded on different tapes. In these instances tape editing requires cutting portions of recorded tapes and splicing them together in the desired sequence. For editing purposes, a recorder is required that allows for the reels to be moved by hand while the electronics reproduce the sounds of the tape. In selecting the editing point, great accuracy for the cut is necessary. In most recorders the mechanical stop is not accurate enough. Once the stop point has been selected, the reels are "inched," moved back and forth by hand, to find the end or beginning of the selected sounds. Once this point is located, the spot of the tape resting on the recording head gap is marked with a grease pencil. Yellow or white pencils are preferred for their visibility.

Next, we need an editing block, a sharp single-edge razor blade, and splicing tape. A splicing block is a metal bar with a slightly curved groove. The groove sides have tiny protruding borders designed to grip the magnetic tape firmly. There are usually two cutting slits across the groove. One is perpendicular to the groove and the other diagonal to it (*see* Figure 7-23A, page 176). The perpendicular splicing cut is used mainly when the tape has voice recorded on it; diagonal splicing is usually done in music editing. This splice will prevent sudden changes in sound level when one piece of the tape ends and the next starts. The change provided by the diagonal splice will be gradual and therefore harder to detect under normal listening conditions. One end of the magnetic tape, with its backing at the top, is pressed into the groove by running a finger lengthwise over it (*see* Figure 7-23B, page 176). The tape in the groove will be held under the edges. It is possible for the tape to slide inside the groove, but it will not move otherwise. This end is then overlapped by the other end, and then, by inserting the razor into the cutting slot, the tape is sliced (*see* Figure 7-23C, page 176). Now butt the two ends together and apply a piece of splicing tape approximately 1 inch long over the tape and burnish it with your finger or fingernail until the splicing tape takes on some of the coloring of the magnetic tape (*see* Figure 7-23D, page 176). The splice is made and the tape is removed from the block by pulling the tape taut from both ends and lifting simultaneously (*see* Figure 7-23E, page 176). *Do not* use a peeling off motion from one end only. This tends to damage the

Figure 7-23. Steps in Splicing Audiotape

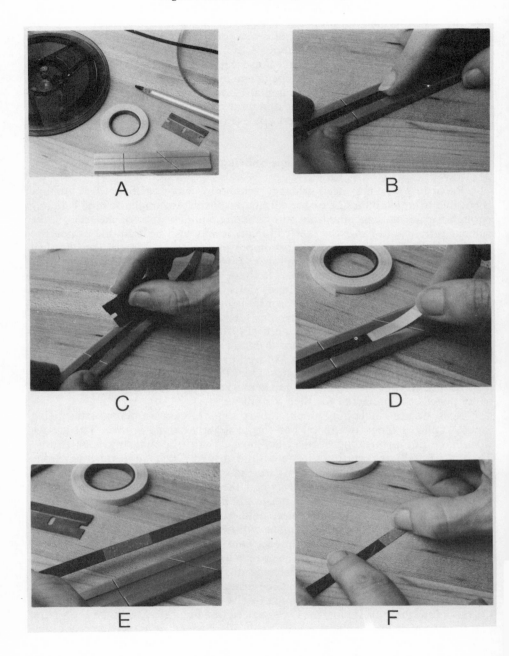

edges of the tape and may cause tracking problems. Also, make sure that the razor is not magnetized. A magnetized razor will leave an "imprint" in the tape that will be heard as a "pop" or a "click" in playback. You may wish to burnish the splice further after the tape is removed from the block to ensure a good bond (*see* Figure 7-23F).

BASIC SOURCES

Nisbett, Alec. *The Technique of the Sound Studio*, 3rd ed. New York: Hastings House, 1972.

Nisbett, Alec. *The Use of Microphones*. New York: Hastings House, 1974.

Schroeder, Don, and Gary Lare. *Audiovisual Equipment and Materials: A Basic Repair and Maintenance Manual*. Metuchen, NJ: Scarecrow Press, 1979.

Tremaine, Howard M. *Audio Cyclopedia*, 2nd ed. Indianapolis, IN: H. W. Sams, 1969.

Woran, John M. *The Recording Studio Handbook*. New York: Sagamore, 1976.

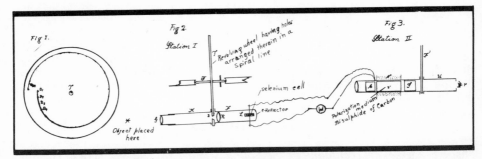

Schematic Drawing of Nipkow's Transmitter Patented in 1884

8 – TELEVISION SYSTEMS

Television, the electronic transmission of moving pictures with accompanying sound, was made possible by a series of discoveries begun in the latter part of the nineteenth century. In 1884, Paul Nipkow of Germany patented a system that led to the transmission of black-and-white moving pictures for the first time in 1925 by J. L. Baird in England and C. F. Jenkins in the United States. This system combined complicated electronic devices. A more efficient system was developed also in 1925, when the American, V. K. Zworkyn, patented the *iconoscope*, a cathode ray tube, derived from a similar discovery by K. F. Braun in 1897. The invention of the iconoscope made possible the television system in use today. The principles of this system were simultaneously suggested in 1907 by a Russian, Boris Rosing, and an Englishmen, A. A. Campbell Swinton.

Television was publicly established in England by 1936. In the United States experimental transmission began during the same year, but official broadcast started in 1941. Until 1956, television broadcasting consisted of "live" action taking place in studios before the cameras or of reproduced motion pictures. In that year the Ampex Corporation introduced the technology necessary to record the television signal. Videorecording made possible the storage of information using magnetic recording principles; playback of the program material could occur later and repeatedly. The process is quite similar to that of producing audio-recordings as discussed earlier.

THE SIGNAL

The television system requires more highly sophisticated equipment for origination and recording than does sound equipment, due to the complexity of the television signal. This signal has two components: the picture information, usually referred to as "video," and the sound information, referred to as "audio." These two signal components are *multiplexed* or electronically mixed for television broadcasting, or they are sent over transmission lines as two simultaneous but individual signals. Therefore, we speak of "television" when both signal elements, audio and video, are mixed and transmitted together, and we speak of "video" when the signals are handled independently. For television viewing, a receiver decodes the multiplexed signal picked up by an attached antenna. On the other hand, a "monitor" is used to receive two cable connections, for video and audio respectively.

THE TELEVISION PICTURE

Before we enter into a description of video cameras and recorders, we should discover how the picture is produced. A cathode ray tube, in which the television picture is reproduced, consists of an electronic beam simultaneously going left to right and up and down inside a vacuum tube. This beam also changes in intensity proportionately to the light and dark values of the picture. The glass vacuum tube is an envelope that contains a screen, made up of minute phosphorus particles arranged in a mosaic-like dot pattern, and an electronic gun or *cathode* (*see* Figure 8-1).

Figure 8-1. Simplified Cathode Ray Tube

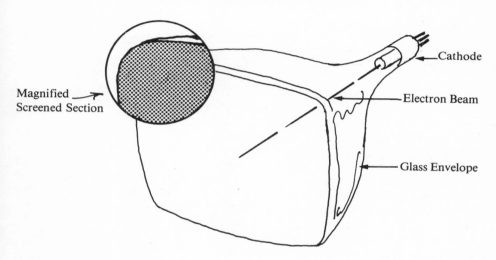

Magnified Screened Section — Cathode — Electron Beam — Glass Envelope

The cathode emits a narrow beam of electrons traveling at a high speed that strikes the phosphorus particles of the screen. When these phosphorus particles are struck by electrons they glow. Television is based on the combination of these events. By controlling the emission of electrons and by selectively choosing which dots are going to light up, an image with dark and light (or color) values is created on the screen. Let's visualize how the television signal controls the functions of the cathode ray tube by comparing it to your vision and eye movement.

As you read this page, your eyes move across each printed line from left to right, and at the end of each line, they move down to the next line until you reach the end of the page, when your eyes will go to the top of another page. The electron beam inside the cathode ray tube functions in the same manner, "scanning" or "tracing" the screen. Besides scanning from left to right, the electron beam also stops emitting electrons at the end of each trace line, and at the bottom of the screen. During this "blank out" time, the beam is repositioned to the starting point of a new trace line by electromagnetic fields of force. This blanking of the electron beam, eliminates our seeing it retrace across the screen (*see* Figure 8-2, page 170). This "start and stop" of the beam would be the equivalent of your closing your eyes at the end of each line so as not to see the page while you reposition your vision at the beginning of the next line or the following page.

Figure 8-2. Video Scanning Lines

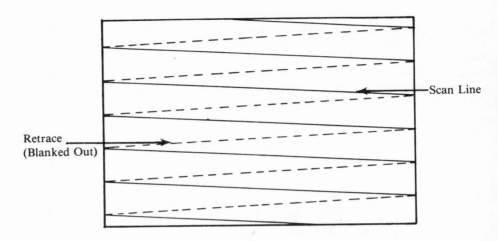

TELEVISION FRAMES

The scanning of the screen area is done in a field with 262.5 lines. But a field is not an entire picture. The scanning employed in present television systems is known as "interlaced" scanning and requires two fields for an entire picture or "frame." Again, let's compare television scanning to our reading method to explain the interlaced scanning system. If we were to begin reading on the top line of the page and then go down to read the third line, and after that the fifth line, and so on, always skipping the even numbered lines, we would be scanning one field (*see* Figure 8-3A). If, after reading the last line on the page, we were to go back up the page and read the even numbered lines, we would have scanned the interlaced field (*see* Figure 8-3B). Now we have scanned two fields or one frame with 262.5 x 2 or 525 lines; the result is a complete picture with the sum of both fields.

Figure 8-3. Interlaced Scanning

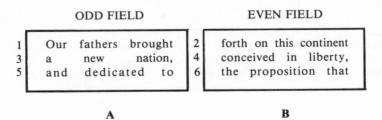

Obviously, this method of reading would not be adequate for our understanding of written information. We read by relating one word to the next in succession. But this kind of scanning is possible for the televised image due to the persistence of our vision, i.e., the ability of our sight to retain an image for a short period of time. Our brain stores the information of the first field, and after the second field is traced, we see the entire frame as if it were scanned all at once. For this to occur these fields must follow in rapid succession, and indeed they do. Sixty fields are traced each second, and since there are two fields per frame, the scan produces 30 frames per second. This scanning speed gives a horizontal line frequency of 15,750 lines per second (60 x 272.5 — field frequency x the number of lines per field). Remember that as the beam moves sideways, it also moves down and then up to begin the new field 60 times per second; this is the vertical frequency. The horizontal and vertical frequencies will provide a standard "synchronization" signal. The signal is used in all components of the television system to maintain all electronic sequences in step with each other or "in sync."

Now that we have scanned the screen of the tube with a beam of electrons, another process also has to be explained. The even tracing of the phosphor screen would give a fully lighted screen. That would be great for a patch of snow on a mountain peak, but not for much else. In order to achieve differences in shading, the video signal has to open and close the cathode gun just as we open and close our eyes, only much faster. It has to stop the electron beam completely for blacks, emit a full charge for whites, and regulate its intensity between these two extremes for grays. As the beam travels, it emits electrons or stops them thousands of times per second. To have a picture with good "resolution" the dots or elements making up the pictures must be very small. The smaller the dots, the greater the detail. And to achieve good sharpness, a screen must be made up of many thousands of tiny dots in close proximity that can be lighted up independently. A television frame contains up to 200,000 picture information elements, which creates an electronic image of good definition. In short, the equipment that reproduces the television image must conform to very stringent electronic standards and must perform well mechanically to carry out its functions.

THE VIDEO CAMERA

In general, the video camera behaves much like a human eye and will convert a scene to the electronic signal necessary to create a video image. A video camera employs for this purpose a lens and camera pickup tube. Earlier, we learned that a microphone uses electromagnetic principles to convert sound waves into an electric current. A video pickup tube, the equivalent of a microphone in video origination, employs *photoconductivity or photoemission* to accomplish similar results. Photoconductivity is the ability of an element to pass an electronic current in reaction to illumination, and photoemission is the property of certain metals to create an electric current also in response to light. A video tube employing photoconductivity controls the flow of an electric current through its face or *target* when light strikes it (*see* Figure 8-4).

Figure 8-4. Simplified Video Camera Tube

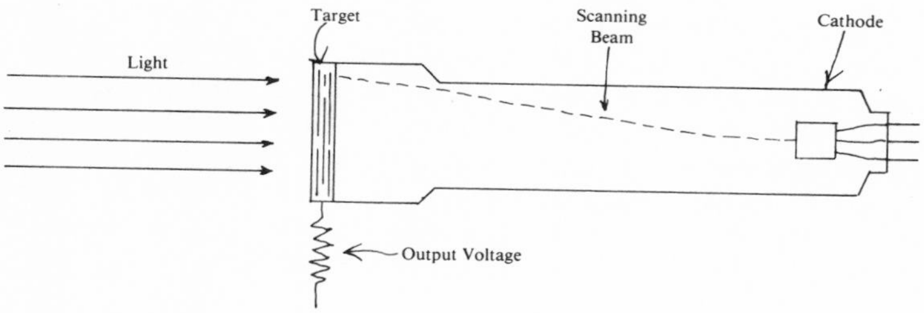

Most portable video cameras have this type of electronic device, called a *vidicon* after the trade name of the first tube of its kind. In contrast, cameras with tubes that use photoemission are larger, but the resulting picture is usually of better quality. Video camera tubes also have an electron beam that interacts with the photosensitive elements to create electronic impulses with the synchronization standards described previously. In this manner, when the camera lens focuses the image onto the face of the video pickup tube, the scanning beam allows current to flow in response to the organized progression that we encountered in the cathode ray tube of the television receiver (*see* Figure 8-5).

Figure 8-5. Video Signal Processing

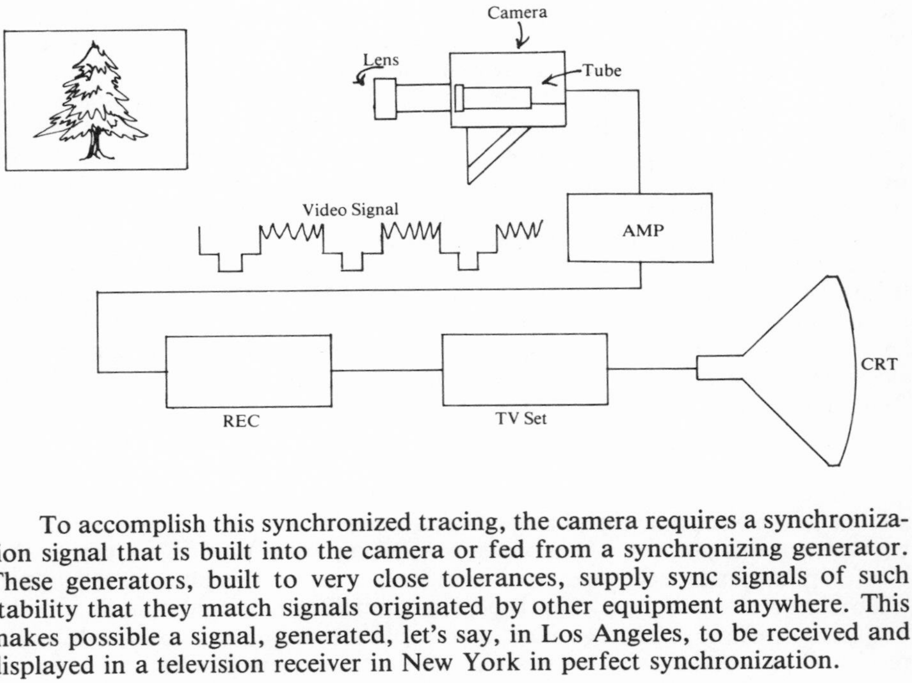

To accomplish this synchronized tracing, the camera requires a synchronization signal that is built into the camera or fed from a synchronizing generator. These generators, built to very close tolerances, supply sync signals of such stability that they match signals originated by other equipment anywhere. This makes possible a signal, generated, let's say, in Los Angeles, to be received and displayed in a television receiver in New York in perfect synchronization.

CAMERA HANDLING

- In handling a video camera care should be taken not to hit it or drop it. The lens is fragile and so is the vidicon. A strong impact may break the glass or misalign the components.

- The camera should never be pointed toward a strong point of light. When shooting outdoors, never point the lens into the sun. This precaution should be observed whether the camera is on or off. Remember that the target of the tube is made of photosensitive material and may be permanently damaged by light. But, it may be possible to eliminate a minor burn. Pointing the camera, while functioning, toward a white, evenly lighted surface for a couple of hours may take care of the problem.

- The camera should not be carried or stored with the lens pointing to the floor. The target of the vidicon can also be damaged in this manner.

- When using the camera on a tripod, lock the tripod tilting mechanism before leaving the camera. Many cameras have been damaged by not paying attention to this simple precaution. Although some cameras are well balanced and may rest on a leveled position, the lens is usually heavy, which will cause the camera to tilt down by its own weight and to hit the tripod. As a result, the lens or internal components of the camera may break.

- Carry a video camera by its handle. The power or signal cable is not a holder and will break or come loose from its connection if it is used as a handle.

VIDEO AMPLIFIERS AND MIXERS

The video signal originated at the camera can be electronically processed either for transmission or recording. Video pickup tubes generate weak electronic currents that are amplified by video amplification stages prior to viewing or recording. If this sounds familiar to you, it should. The audio signal processing explained earlier is similar to video processing as discussed this far (compare Figure 8-6 with Figure 8-5). The difference begins at the mixing stage. When making a videorecording with several cameras, the purpose is to select one camera angle from different views or scenes that may be happening at the same time. You remember that in audiorecording we don't select microphone signals but mix and add them, seeking to maintain the continuity of speech or music. In video productions one "shot" replaces another. The video mixer is a "switcher"; it selects a single camera picture (*see* Figure 8-6).

Figure 8-6. Video Switcher

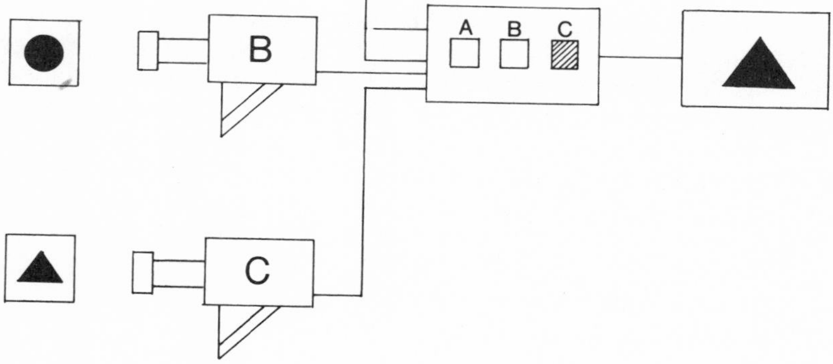

Only for special effects, such as fading from one picture to another or "inserting" one camera shot partially over a second camera shot (split screen), do the switches add up signals from more than one video source (*see* Figure 8-7).

Figure 8-7. Special Effects

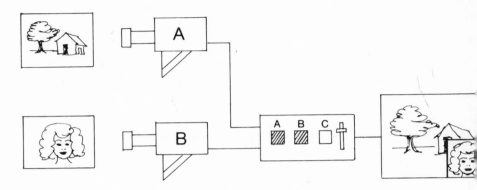

This equipment has special effect generators that allow for a variety of video screen shapes and combinations of fades or split screens.

CABLES AND CONNECTORS

Video equipment and components require more than one signal. There are different types of cables and connectors with single and multi-pin hookups that usually match equipment from the same manufacturer. In hooking these cables, care should be given to the purpose and the requirements of each component. To prevent damage, be sure equipment and components are compatible prior to hooking them up. Video connecting cables that carry one signal are usually of a coaxial type with a standard UHF connector. These signal lines have an impedance of 75 Ω . Other cables have several signal lines inside of a single jacket (*see* Figure 8-8). The most common multi-signal cable has an 8-pin connector with two of the pins spaced differently from the other six to prevent changing the orientation of the plug.

VIDEO RECORDERS

Video recorders are truly remarkable feats of engineering. These units perform very complex and accurate tasks and still are relatively simple to operate and maintain. The electronic complexity of the television signal demands a great deal from video recorders. Among other requirements, these recorders must:

- record signals of high frequency,
- record audio and synchronization signals simultaneously,
- handle tape gently at high speeds,
- maintain tape speed constant, and
- provide an accurate tape path.

Figure 8-8. Video Cable and UHF Connector

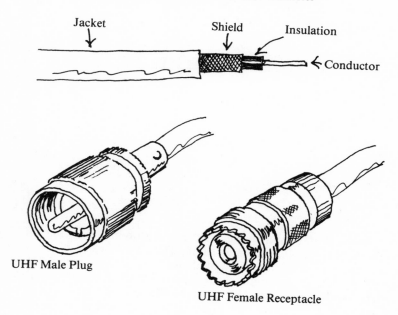

Jacket   Shield   Insulation

←Conductor

UHF Male Plug

UHF Female Receptacle

Recording Formats

Early attempts at magnetically recording the television signal were done in equipment similar to audiorecorders. This type of recording is known as "longitudinal" recording. That is, the tape travels past a stationary recording head providing a continuous recorded track lengthwise on the tape (*see* Figure 8-9A, page 186). Video quality in these recordings was poor, mainly due to tape speed. Also, the amount of tape needed for a good recording was excessive. We recall that to record high audio frequencies faithfully we used a high tape speed. Audio frequencies below 20 thousand are low in contrast to the video frequencies ranging above 4 MHZ (MHZ = megahertz = million cycles per second). Tape speeds necessary to record these frequencies would require a few thousand feet of tape for a minute of video programming. The solution to the impractical and costly longitudinal recording for television was introduced by the Ampex Corporation with the "transverse" recording format (*see* Figure 8-9B, page 186). By having a video head assembly rotating perpendicularly to the tape direction the "video writing" speed was increased, which utilizes less tape and reaches effective high frequency recording.

This system has four video heads placed at right angles on the edge of a drum (*see* Figure 8-9C, page 186). The head drum assembly rotates at a high speed. The heads come in contact with the tape traveling on a path perpendicular to the motion of the drum. These combined motions—the heads rotating at a high speed in one direction, and the tape moving at a slower speed in a different direction—result in a head-to-tape speed or writing speed that is adequate to handle the requirements of high frequency recording for the television signal. This transverse recording format is known as *quadruplex* or *quad* for short. It is used in high quality production equipment and employs 2-inch wide magnetic tape. In

Figure 8-9. Videorecording Tracks

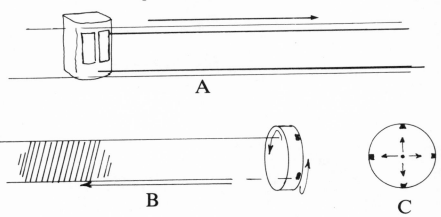

our work, we will employ a simpler and more economical format of excellent quality: the *helical* or *slant track* recording. The slant track recorder can be considered a combination of the longitudinal and transverse recording formats. The tape travels much as in an audiorecorder and the heads rotate on the same plane, but the tape is slightly slanted in its path. These recorders usually have two videorecording heads located 180° apart on a drum assembly. The tape wraps around the drum on a *helix*—higher at one point, lower at another—as it travels on its path (*see* Figure 8-10). Helical recording therefore induces video tracks at a slant greater than those on a "quad" recorder. These recorders use ½- or 1-inch magnetic tapes.

Figure 8-10. Helical Videorecording

Tape Interchange

We indicated in the introduction (and in the illustrations) that a videorecorder must record simultaneously video, audio, and synchronization signals. Videotaping a classroom scene or a dramatic performance would be almost useless if only the image were to be recorded. The sound accompanying the action must also be recorded to provide an accurate account of the activities.

For this purpose a videorecorder has audiorecording capabilities as well. The nature of the television system of line recording and subsequently the videorecording systems just described, requires that the video heads "read" each recorded track precisely. The heads must always track each line of recorded information discriminating from adjacent lines. To accomplish this, a control track is also recorded. This control track adjusts the tape speed so that the prerecorded video tracks can be "read" by the video heads accurately. The space between the video tracks is measured in thousands of an inch. Another important fact is that if a recorder uses one recording format, a tape recorded on that unit cannot play back on a recorder using a different format. As illustrated in Figure 8-11, it is apparent that one tape cannot be exchanged for the other. Earlier manufacturers of recording machines used their own tape format layouts, recording speeds, tape sizes, etc. A tape recorded in, let us say, a Sony ½-inch recorder could not play on a Panasonic ½-inch recorder, or vice versa. Standards established by the Electronic Industry Association (EIA) now make tape interchange possible. We encounter in our work more frequently the format designated as ½-inch EIA-J. If you find a tape that does not reproduce on a video player that operates well with other tapes, it is likely that the recording format is not standard and tape interchange is not possible. Tape interchange refers also to a recorder's capability to play back tapes recorded in another machine of the same format. All machines are manufactured in compliance with the standardized formulas, but machine use, handling, and cleaning will affect parts alignment. Normal wear of guide posts, friction rollers, capstan, etc., will also affect interchange.

Figure 8-11. Different Videorecording Formats

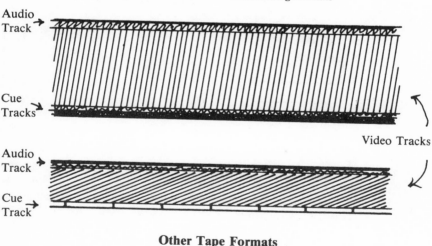

Audio Track →

Cue Tracks →

Video Tracks

Audio Track →

Cue Track →

Other Tape Formats

So far we have talked about videorecorders of open reel design. Just as in audiorecorders of this type, the tape travels from one open reel to another. In this type of machine we discussed the 2-inch quadrature format, and the ½-inch helical EIA-J format. Others, some old, some new, employ ¼-inch tape and 1-inch tape. Videocassette or cartridge recorders are becoming rather common in learning centers, libraries, and media production facilities. Among the many available recorders you may find the U-matic cassette, and the

V.H.S. formats. U-matic recorders and players use a ¾-inch tape contained on a cassette allowing up to 60 minutes of playing time. The V.H.S. format uses a ½-inch tape cassette and provides up to 120 minutes of continuous playing time. Both formats are standardized by the EIA, and there are units for both types that can provide extended playing time (120 and 240 minutes respectively). Note: unlike audiorecording tape, in reels or cassettes, which can be used in both directions, video is recorded in only one direction due to the format layout.

Care of Videorecorders

Videorecorders need to be cleaned more often than audiorecorders. The video head, due to its size and its speed, comes in contact with the tape more than the head in an audiorecorder recording for the same running time. The head also has a smaller gap and is more susceptible to "clogging" with oxide. A magnetic particle flaking off the tape can close the video head gap preventing recording or playback. Head cleaning should be done after every hour of use and with great care. The video head is very small and brittle. The cotton swab must be passed over the head sideways *not* in an up and down motion.

Figure 8-12. Video Head Cleaning

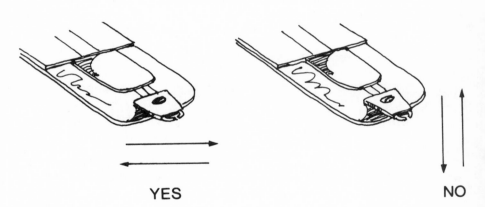

YES   NO

Up and down rubbing may chip the head, damaging it permanently. Since the head assembly rotates, it is necessary to hold the head while cleaning. The head is placed on the center of a metal bar. You hold this bar in place by pressing lightly with the end of a cotton swab, while cleaning the head itself with another swab dipped in cleaning solution. Be careful not to press on the head by mistake.

Make sure that no lint is left on the head from the swab. Magnetic head cleaning fluid in spray cans eliminates the danger of damaging the tape, but in excessively dirty equipment it will not do the job well. Practice cleaning a video head, and if possible, watch someone with experience before you attempt it. All guide posts, rollers, and any other parts that touch the tape also must be free of oxide. Remember that a slight misalignment on the tape travel will result in a head's mistracking. Be especially careful with the surfaces touching the tape at its low edge since this is the "guiding edge." Never rest your hands or place any heavy object on top of the head drum assembly. This could warp the head assembly, which will also result in mistracking conditions.

Mechanical equipment with motors, lamps, or electronic tubes generate excessive heat that will damage components even though most recorders have air vents to help keep interior parts cool. Care must be taken not to place units on cushion chairs, or otherwise obstruct free circulation of air through these vents.

Demagnetize heads and other metal parts often, and observe care in handling the degaussing instrument. Do not hit the video head with it.

In threading the videotape onto the take-up spool, do not fold or crimp the ends. Let the tape itself overlap the end flatly by revolving the reel. A crimped or folded tape will stretch it, which is even more critical in videorecorders than in audio. The control track and the video tracks are spaced with close tolerances. The slightest shift will result again in mistracking. Loss of control pulse tracking will be visible as a vertical rolling of the picture. Mistracking of video tracks is visible as horizontal "tear" or "noise" in the screen.

VIDEOTAPE SPLICING

Videotape splicing is done only to repair a broken or cut tape; it is not done for editing. Editing is done electronically. That is by copying or rerecording from one tape onto another in the desired sequence. Splicing videotape is not recommended. To splice videotape properly requires a long, precise, transverse cut on the tape. To make a splice not visible on the screen the tape has to be slit between two video tracks (which are illustrated in Figure 8-11, page 187). For example, the length of the track in the EIA-J format is 7½ inches. The spacing between tracks is approximately 1 mil, and the track width is approximately 1.7 mils. To make a splice the tape must first be wetted in a "developing" solution, a liquid that allows the magnetic orientation of the oxide particles to become visible with the unaided eye. An editing block similar to the audio splicing block (*see* Figure 7-21, page 173) is used. The tape is slit between two video tracks and joined with a mylar tape. Due to its length, the cut is also more susceptible to dirt accumulation. Dirt acts as an abrasive on the heads. It is preferable, when affordable, to replace a videotape rather than to repair or splice it.

SUMMARY OF TELEVISION

1) A video camera converts a scene into an electronic signal using photographic and magnetic principles.

2) A video signal fed into a cathode ray tube reproduces the scene using the same principles.

3) Television, through complex and sophisticated technology, combines a video and an audio signal to reproduce a true account of a scene or event.

4) The presently used television system is possible largely due to the persistence of human vision.

5) Videorecording and video processing equipment allow for storage of visual and sound information.

BASIC SOURCES

MacRae, D., M. Monty, and D. Worling. *Television Production: An Introduction.* Toronto: Metheuen, 1973.

Robinson, Richard. *The Video Primer.* New York: Links, 1974.

Williams, Richard L. *Television Production: A Vocational Approach.* Salt Lake City, UT: Vision Inc., 1976.

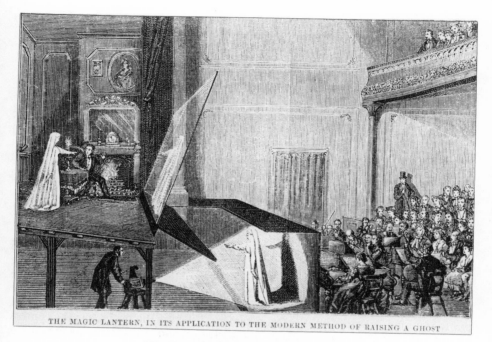

THE MAGIC LANTERN, IN ITS APPLICATION TO THE MODERN METHOD OF RAISING A GHOST

The Magic Lantern in Public Use
(Reproduced courtesy of the Gernsheim Collection, Humanities Research Center,
The University of Texas at Austin.)

9 – PROJECTION EQUIPMENT AND MATERIALS

PROJECTION OF IMAGES

Early conjurers and magicians amazed and amused audiences with projection techniques that opened the way for the creation of a vast array of devices for the projection of still images, which culminated with the development of motion pictures. By the middle of the seventeenth century, Athanasius Kircher, a German Jesuit, created the *magic lantern*. This projection device used a light source illuminating an image painted with transparent colors on a thin glass square. The image was cast on a flat surface with the aid of a lens. An optician and magician, Robertson (Etienne Gaspard Robert) perfected this projector and used it publicly for the first time in 1784.

The Basic Projector

Kircher's invention provided the basic ideas for the design of the projectors in use today. The main components are: 1) a source of light, the projector lamp, which illuminates 2) a transparent image that is focused by 3) a lens on a flat surface (*see* Figure 9-1, page 192).

Figure 9-1. Basic Components of a Projector

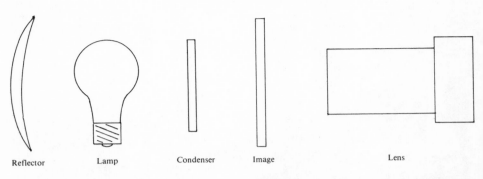

Reflector   Lamp   Condenser   Image   Lens

In addition, a projector may have a reflector to direct the lamp's light and condensers (insulating lenses) to prevent the image from burning under the intense heat of the light source. The lens is usually movable and permits the focusing of the image on the screen. In chapter 6 we learned that focusing the camera lens provides a sharp image on the film. Projecting is the opposite of photographing: whereas light enters a camera, it leaves the projector; the camera lens focuses the image onto the film, while the projector focuses the image onto a screen or viewing surface.

Lantern Slide Projectors

Lantern slide projectors are still in use in many classrooms, and don't be surprised if you find commercially made glass slides in black and white or even in color. The slides of this modern version of the magic lantern measure approximately 3¼x4¼ inches and have taped edges to prevent cut fingers from the glass. Instructional kits containing glass rectangles, protecting tape, and color grease pencils used for writing on the glass surface are available. These kits are used by teachers and pupils alike for making their own visuals. A similar slide format was also introduced by the Polaroid Corporation as part of their instant film line. These slides are made with B&W transparency film and are mounted on plastic carrier frames. Although this system produces slides that are more expensive than handmade slides because of the necessary equipment—i.e., a Polaroid Land Camera—it has the advantage of making large slides of good quality in a matter of seconds. The films used—Polaroid Land pack types 46 (continuous tone) and 146L (high contrast)—require 120 seconds and 30 seconds of development time respectively, and the fixing of the image to the acetate base takes 15 seconds. The fixing chemical is provided in a container that doubles as a fixing tank. After the film is exposed in the camera and subsequent development has occurred, the slide film is inserted through the rubber slit at the top of the container holding the fixing solution. This container is locked by a plastic clip, and agitated. The film is then pulled out through the rubber slit that acts now as a squeegee due to the pressure of the locking clip. The slide dries in a matter of seconds at room temperature. When dried, it is inserted in a snap-on mount and is ready for projection. The use of this type of slide might not have developed widely for several reasons: 1) this instant film is only available in black and white, 2) the required Polaroid Land camera is available with specialized equipment and

found only where its application warrants the cost, 3) perhaps the more important reason, use of the 35mm or 2x2-inch slide and the Kodak carousel projector
is widespread.

2x2-inch Slide Projectors

35mm reversal film gives us color transparencies of excellent quality at
affordable prices. This film can be exposed in any 35mm camera, and commercial
processing is readily available. The film frames are returned by the laboratory
mounted on 2x2-inch thin cardboard mounts. To project these slides a wide variety of projectors are available, from the simplest hand-operated "mini" projector
resembling a midget magic lantern, to the leader in the market, the carousel projector (*see* Figure 9-2).

Figure 9-2. Carousel Projector and Tray

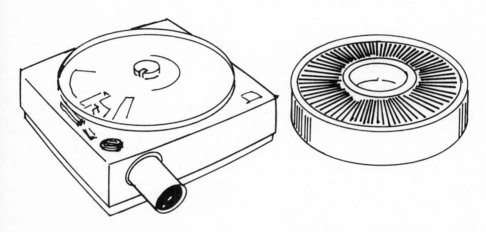

The carousel projector has a round tray in which the slides are loaded. This
tray resembles a wheel with slots lined up as spokes into which the slides are
inserted. The carousel tray sits on top of the projector, centered on a round post.
As the projector's *advance* button is operated the slide on the slot resting directly
over the projection gate drops into it. When the advance is operated again, a
combination of levers and pulleys performs a series of actions: 1) the slide on the
gate is pushed up into the tray, 2) the tray is advanced, and 3) the next slide in the
sequence is dropped into the projection gate. This ingenious mechanism can go
on, and on, with the tray revolving around the center post, either forward or
backward. Carousel trays hold 80 or 140 slides. Regardless of their capacity, they
are of the same circumference and are interchangeable; therefore, to fit more
slides on a tray requires thinner slots. The thickness of the slide mount determines
the type of tray to use. Cardboard mounts and some thin plastic mounts can be
used in a 140-slide tray. Glass or heavy duty mounts must be used with an 80-slide
tray.

SELECTION AND CARE OF SLIDE MOUNTS

Cardboard mounts provided by processing laboratories are good in most cases. However they are fragile and may warp, and after heavy use, they fray at the corners. The carousel projector is gravity-fed, i.e., the slide's own weight makes it fall into projection position. A defective slide mount will either drop partially into the gate or will not drop at all. If the mount corners are slightly frayed, you may repair the mount. A simple and quick way to mend it is to clip the frayed corner slightly and wet the cut edge with white glue. If the corner is only beginning to fray it may suffice to dip it in white glue and, as the glue turns "tacky," to squeeze it flattening the frayed corner. White glue dries hard, preventing repeated fraying.

When the mount is damaged beyond repair, you should place the slide on a new mount. Slides are also remounted in better mounts to protect and preserve them. A good variety of slide mounts that fit any purpose and budget are available. Some cardboard mounts can be sealed at the edges by applying heat with a tacking iron. (This instrument and dry mounting techniques are explained in chapter 10—*see* Figure 10-6B, page 228). These mounts are easy to replace and are inexpensive. Unfortunately, they are subject to the same wear that required remounting of your slides in the first place. They should be chosen only for slides with expected little use or for loading 140-slide trays. Many slides also have rounded corners, which should provide longer wear without fraying. A bit more expensive are thin plastic mounts, such as those produced by Lindia or Kindermann, that can also be inserted in 140-slide trays. These mounts are also sealed by heat and usually require a specially designed sealing unit. The necessary hand-operated mounting units are rather inexpensive, but the quality of these mounts varies greatly. The mounts are also susceptible to warping by the heat generated by the projector's lamp. Slides that are projected for longer periods of time in poorly ventilated areas should not be mounted on thin plastic mounts.

The better mounts are thicker, and as you might expect, more expensive per unit. They are preferred for many reasons. These mounts will not need replacement under normal use and wear; only severe handling will damage them. They will not be warped by normal projector heat even if projected for unusually long periods. Their heavier weight will help them drop better into projection position in gravity-fed projectors. These mounts are also available with glass covering the frame aperture. Glass will protect the film from dust and fingerprint marks. This glass can be made of plain or Anti-Newton-ring glass. Newton rings appear during projection and are concentric rainbow colored rings. They are produced by heat and condensation trapped between the film and the glass. Anti-Newton-ring glass prevents the formation of these rings that shift over the surface of the transparency as it changes temperature. These good quality mounts are usually of a snap-on type and do not require special sealing equipment. Slight finger pressure will lock the mount halves together.

REMOUNTING SLIDES

Remounting slides requires removing the transparency from the laboratory provided mounts. Work on a surface that is free of lint and dust. With a sharp blade, such as an x-acto® knife or a single-edge razor blade, split the mount apart at one corner (*see* Figure 9-3A). Tear the mount apart by pulling from the two split ends (*see* Figure 9-3B). More likely you will expose at least one edge of the transparency, and if the cut was right at the junction of both halves, you may expose the entire transparency. Now, you may peel off the transparency with care. The film is heat-sealed against the mount at a spot (usually the bottom sprocketed edge—*see* Figure 9-3C), and the bond can be broken by pulling from the transparency slightly harder.

Figure 9-3. Opening a Cardboard Slide Mount

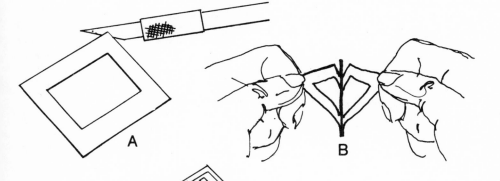

The transparency should be dusted before it is sealed in a new glass mount, or after it is sealed in a glassless mount. For dusting, use a camel hair brush. Brushing the film will create a static charge, especially in dry weather, that will attract dust to the brush. Clean the brush frequently, and dust with continuous strokes over the transparency, otherwise you may apply dust instead of removing it. Antistatic brushes are available, and it is best to obtain one of these types for the job. Slide mounts hold the transparency in place either by providing a recessed rectangular area of the size of the transparency frame, or by sliding the film edges under tabs or guides. One of the best mounts, made by Wess Plastics, has small posts that fit the film sprocket holes securing the transparency in place. These mounts are made to strict standards and are used in professional work requiring exact registration of the slide.

Slide Positioning

In mounting slides, care should be given to orient the emulsion side of the film toward the lens side of the projector when the slide is in projection position. Some projectors have automatic focusing mechanisms. These projectors adjust focus by measuring the distance of the film surface from the lens' optical element.

The emulsion surface provides a better measuring reference point. Reversing slides in their mounts will provide erroneous measuring for the automatic mechanism resulting in out-of-focus projection. The emulsion surface of the film has a matte look in some areas. The photographic image is "etched" on the surface and appears rougher to the eye than does the film's celluloid backing, which is glossier and smoother.

Positioning projecting materials for "right-side up" viewing is a habit you should develop. There is nothing more annoying to an audience than to have visuals presented upside down or reversed. It is good for an initial laugh, but it is a situation that can be avoided with small effort, and here is how it should be done. Optical systems, lenses, and eyes, shift the light rays carrying the visual information. That is, point A on the slide in Figure 9-4 appears at the upper right corner. After pasing through the lens it will be the lower left corner of the projected image.

Figure 9-4. Reversing of an Image by a Lens

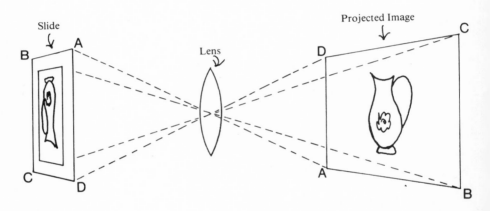

Figure 9-5 indicates the positioning of a slide for projection. Hold the frame so that the image is upside down and reversed. Remember that all projection materials have to be loaded onto a projector in this manner. Slide mounts that are returned from processing laboratories usually have a printed number on a corner corresponding to the negative number on the film roll sequence (*see* Figure 9-6). This number is leaded on a projector so that it can be read properly if we look at it from the projector lamp side. There is one exception however; if the slide is in a vertical format the number will be shifted to the upper left corner.

Figure 9-5. Slide Positioning for Projection

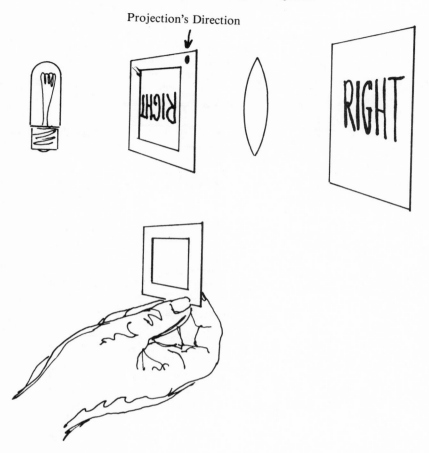

Projection's Direction

Figure 9-6. Vertical Slide Positioning

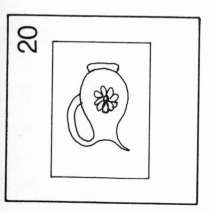

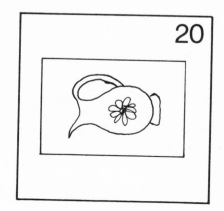

FILMSTRIPS

Another variety of material for projection commonly found in education is the filmstrip. A filmstrip is a series of photographic frames on an uncut length of 35mm film. The 35mm reversal film used to make slides can also be loaded in a *half frame* camera to produce filmstrips. The 35mm camera exposes film with frames running lengthwise along the sprocket holes (*see* Figure 9-7).

Figure 9-7. 35mm Full Frame Filmstrip

In contrast, a half frame 35mm camera exposes frames of half the size with the horizontal dimension perpendicular to the sprocket holes (*see* Figure 9-8).

Figure 9-8. 35mm Half Frame Filmstrip

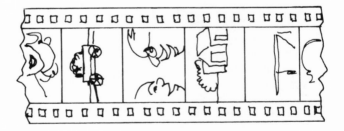

Therefore, on the same length of film, twice as many frames can be obtained with a half frame camera. A 20-exposure roll provides a 40-frame strip, and a 36-exposure roll will provide a 72-frame strip. The standard for filmstrip production requires a "leader" of at least 12 frames and a "trailer" of usually four to six frames. A film leader is necessary to thread the film through the projector onto the film take-up reel. The beginning of the film must have a length of film without information so that when the film engages the take-up reel, the first useful frame is in the projection gate ready for projection (*see* Figure 9-9).

The trailer is a length of film long enough to protect the outer surface of the filmstrip when coiled for storage. The length is determined by the circumference of the packing container. This layer of film protects the filmstrip from scratches that may result while putting in or taking out the filmstrip from its storage container.

The number of remaining information frames on a strip shot with a 36-exposure film roll is then reduced to approximately 54 frames (72 frames less

Figure 9-9. Filmstrip in Projection Position

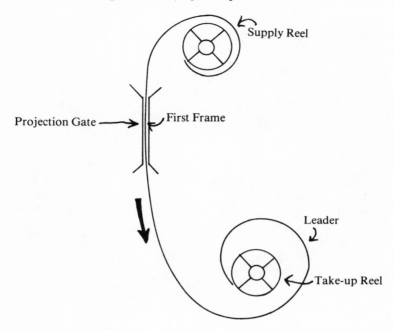

12 for loading and 4 or 6 for the trailer). To make a filmstrip with a standard roll of film requires careful planning. A bad shot in the sequence would ruin the entire roll. Usually a filmstrip is produced by copying individual slides previously made of the subject that will make up the finished filmstrip sequence. (Filmstrips can also be made by photographing art work or pictures made to a predetermined size. A 35mm camera and copystand are required for this procedure.) Why go to the added expense and care that a filmstrip requires when slides have to be shot anyway? The answer is that a filmstrip presents and preserves frames together in a predetermined sequence. With a filmstrip, the possibility of misplacing one transparency of a set or of showing them in the incorrect sequence does not exist. In cases in which the user may not be familiar with the material and equipment or in which a youngster is studying independently of the teacher's supervision, a filmstrip is preferable to loose slides.

Filmstrip Projectors

Filmstrip projection equipment is readily available in instructional facilities. Many manually operated slide projectors can be adapted to show filmstrips and vice versa. Even Kodak carousel slide projectors can be fitted with a filmstrip adaptor; this adaptor has a filmstrip carrier behind a fixed lens that replaces the normal projection lens (*see* Figure 9-10, page 200).

Figure 9-10. Filmstrip Projector and Lens Adaptor

Handling Filmstrips

Since filmstrips are not bound onto reels or spools like other film or tape materials, in handling a filmstrip, avoid placing your fingers on the filmstrip surface. Handle the strip by the edges to prevent fingerprint marks. Filmstrips are stored in small round containers that protect them from dust and damage. Place the strip in the container coiled along the natural curl of the film acetate base. This curling facilitates handling and loading the filmstrip on the projector.

STILL PROJECTION WITH SYNCHRONIZED SOUND

Filmstrips and slide sets are commonly combined with recorded narrative. These multimedia instructional kits are distributed widely, and you will use them often. Early multimedia kits consisted of filmstrips with phonograph records and a teacher's manual, but since audiocassette recording and reproducing equipment was introduced to the classroom, commercially available filmstrips and slide sets are now supplied with the sound on audiocassette. To present photographic images simultaneously with audio narration requires a system to synchronize the media. When the visual information is projected on the screen the corresponding narrative should be heard over the loudspeakers of the audio equipment. Phonograph records that accompany filmstrips have a "beep" on the audio channel that indicates when the next frame on the sequence should be shown. The "beep" of signal pulse is usually annoying to the listener, but it was a necessary condition prior to the advent of automatic advance equipment that is activated by an inaudible pulse. The phonograph record made for synchronized projection has a system identified as the 30/50 standard. There is a 50-cycle tone recorded as a constant signal that prevents the automatic mechanism from being activated by stray mechanical or electronic vibrations. A 30-cycle pulse is recorded at the point where the frame advance must occur. Both signals are electronically filtered, or separated, from the audible frequency range. Obviously, to use this material without the help of a projectionist to change the frames requires the use of the appropriate automatic equipment. Since this equipment is not always available, the phonograph record may have on one side the inaudible projector activating frequency, and on the other the familiar audible "beep" which can be a short burst of a 400 Hz, a chime, a buzz, or another easily identifiable sound among the narration or musical background of the presentation. The recordings that are provided with audiovisual kits are presently not fully standardized, although more and more production companies are adopting the recommended standards of the American National Standards Institute (ANSI). These standards call for a set of electronic frequencies assigned to different functions.

The 50 HZ Superimposed System

The 50 Hz superimposed system superimposes the 50 Hz pulse over the program material. Similar to the phonograph records, all frequencies below 120 Hz are filtered out of the program to provide a "clear channel" for cue or advance pulse of 50 Hz. In this system the cassette tracks recording configuration will be as indicated in Figure 9-11.

Figure 9-11. Superimposed Audio Synchronization Format

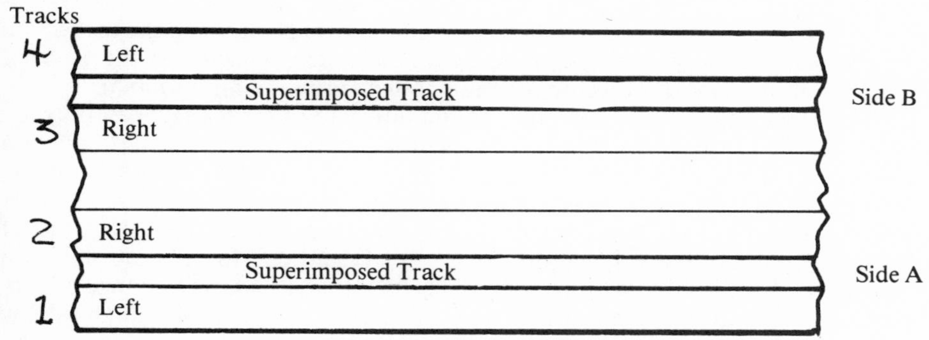

Separate Track System

Another system utilizes separate tracks to record the synchronization pulses. In this format the control pulses can be recorded and erased independently of the program content audio signal. Having separate tracks for synchronization signals allows space for recording other cue signals for other program functions besides "advance." The "advance" frequency allocated is 100Hz. A 150Hz signal is allocated for automatic stop. Two more frequencies have been assigned for future control purposes: 400Hz and 2,300Hz (*see* Figure 9-12, page 202).

Figure 9-12. Separate Track Audio Synchronization Format

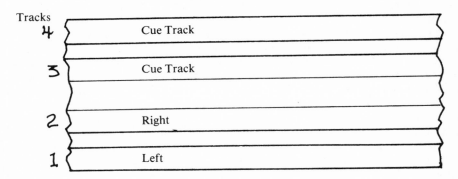

Tracks

4   Cue Track

3   Cue Track

2   Right

1   Left

Synchronizing Equipment

The great variety of equipment and different systems indicates that you must match the right material with the right equipment. Due to the physical configuration of the tracks, cassettes with the 50Hz system cannot be interchanged with cassettes recorded with the separate track system. The recording format should be identified prior to obtaining projection equipment. To add to the variety of self-contained projection equipment that plays back audio and projects the image, some projectors accept synchronization signals from independent audio playback equipment. Here again, the recording format has to be matched to the advance mechanism of the projector. In this case, it is also necessary to employ synchronization cables that will interlock the audio playback unit and the projection equipment. The advance cue signal output has to be connected to the advance input connector in the projector for the unattended presentation mode.

In educational institutions, slide/sound units are familiar gear. Not only are they available for playback, but several manufacturers of audiovisual production equipment offer recorders that permit recording of synchronized programming effortlessly. There are simple units controlling one projector and highly sophisticated units that control up to 24 projection units simultaneously. Units of this latter type often use synchronization signals and control mechanisms with computerized memories and functions. They do not allow for interchange of program materials and are often costly. They also can be interfaced with high fidelity audio equipment suitable for large audiences. Among the simplest units is a synchronizer made by the Kodak Corporation that allows the synchronization of a carousel projector and a reel-to-reel stereo recorder. This unit can help to produce a slide/sound presentation inexpensively, with readily available equipment. It is fair to expect that even the smallest center has a carousel projector and a reel-to-reel stereo tape recorder. The Kodak AV Synchronizer, as it is called, is a self-contained oscillator that generates an electrical pulse signal. This signal is recorded on one channel of the tape simply by connecting one cable to the input of the audiorecorder and depressing the advance of the projector that is connected to the synchronizer by its second and last cable. Once all control pulses are recorded, the connection to the recorder is shifted from the input to the output of the same recording channel and the tape is rewound. During playback the prerecorded pulses will advance the projector automatically and, of course, inaudibly. It is a simple, but effective way to make an inexpensive slide/sound presentation (*see* Figure 9-13).

Figure 9-13. Hookup of a Kodak AV Synchronizer

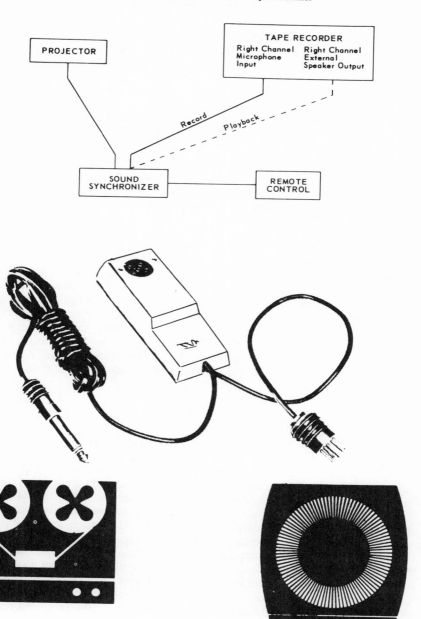

From the array of sophisticated equipment designed for presentation of slide/sound materials providing all sorts of visual effects, we should single out another, more simple but effective unit that controls at least two projectors simultaneously in a basic function: "crossfading" or "dissolving." Two projectors aimed at one screen with overlapping images can project an image on the screen at all times without the "black out" interruption that occurs when a slide is changed. On command, a dissolver will dim the light of one projector gradually while the light of the second projector comes on at the same rate. One projector will be off when the other is on. The transparencies will be shown alternately, fading in and out: slide changes will occur when the projector lamp is off. The advance can be activated manually or by the synchronizing methods described previously.

MOTION PICTURES

Inventors, scientists, and researchers, in an endeavor to step beyond the known or given, employ their curiosity and inventiveness to look for new ways of improving old systems. The magic lantern, early photography, the study of human vision, and recording techniques served their own purposes in isolation. Then came individuals like Sellers, Ducos du Hauron, Baron Uchatius, Muybridge, Edison, and others, who created a series of devices with such names as the Zoetrope, Phasmatrope, Kinetoscope, or Eidoloscope. These inventions attempted to simulate motion by the use of mechanical, electrical, and/or optical principles and mechanisms. In 1895, the Lumiere brothers of France with their Cinematograph, and Thomas Armat of the United States with his Vitascope independently introduced inventions that led to the development of the motion picture equipment in use today. The first appearances of the motion picture were in vaudeville or magic shows, and soon these strips of silent projection material captivated audiences the world over. In 1927 the first "talkies" — motion pictures with sound — further established motion pictures as the most popular form of mass entertainment. The appeal and the advantages of movies are also very applicable for training purposes.

Screen Motion

The illusion of movement on the screen produced by the projection of a series of still pictures is possible because of the persistence of human vision. In chapter 7 we introduced this phenomenon — our eyes maintain an image for a few moments after the image has disappeared. In television we employ visual persistence to scan two fields, one after the other, and to make them appear as if traced in one scanning sequence. In motion pictures we use this principle with a slight modification. A sequence of images with slight variations in each image, presented to us in rapid succession, will be perceived as one image in movement. The movie camera photographs a series of still pictures of an object in movement. Each movie frame consists of an individual image of the object at different points during its movement through space. These frames are recorded with a movie camera at a rate of 24 times in one second. For instance, if a ball rolls from point A to point B in a second, a standard, sound movie camera would record that movement in 24 frames (*see* Figure 9-14).

Figure 9-14. Screen Motion

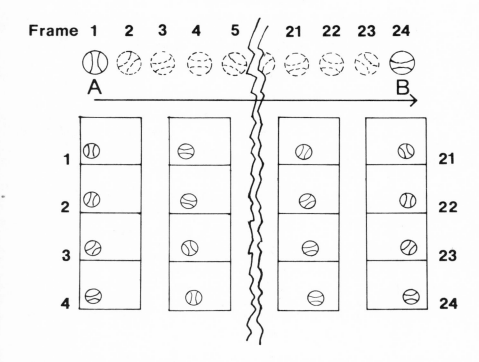

This gives a sequence of 24 pictures in which the ball appears at a different place on each successive film frame. Projecting these 24 frames in one second will let us see the ball rolling across the screen. With this seemingly simple principle, complex movements can be simulated and even dissected for study.

Special Effects

As noted above, movie cameras have mechanical or electrical motors that maintain a constant running speed of 24 frames per second for normal action filming. Altering the speed of the motor, making it run faster or slower, permits the creation of techniques such as slow motion, fast motion, and animation, which can be employed effectively in instructional film sequences.

Slow Motion

Continuing with our example of the rolling ball, let us suppose that the ball was photographed while the camera was made to run faster. If 120 frames of film were exposed in one second, and the projection speed was kept at the standard 24 frames per second, what would be the result? We would see the ball traveling very slowly, taking five seconds to move from point A to point B

$$\frac{120 \text{ frames/sec.}}{24 \text{ frames/sec.}} = 5 \text{ seconds}$$

If the camera speed were increased to 240 frames per second, the projected screen motion would show the ball taking 10 seconds to roll from point A to point B. The movie camera effectively slows down the movement and the film allows us to observe what our eyes cannot detect. In this way, we may study the swing of a baseball bat, the leg movement of a running horse, the fragments of an object being broken, or even a bullet's flight through space.

Fast Motion

The contrary will happen if the speed of the camera is reduced during filming. If our familiar rolling ball is photographed six times in one second, during normal projection speed, the ball would take now ¼ of a second to travel the prescribed space

$$\frac{6 \text{ frames/sec.}}{24 \text{ frames/sec.}} = 0.25 \text{ second}$$

The movement will be so fast, it will be almost invisible to our eyes. But let us move away from our rolling ball and think about a blossoming flower. If this flower takes 12 hours to open from a bud to a full blossom and the camera photographs one frame every 30 seconds, during projection at normal speed we will see the flower blooming in one minute

$$\frac{2 \text{ frames/minute x 60 minutes/hour x 12 hours} = 1440 \text{ exposed frames}}{24 \text{ frames/second projection}} = 60 \text{ sec.}$$

Employing slow motion film expands time; with fast motion, film effectively compresses time.

Animation

The capacity of the movie camera to speed up or slow down action, combined with its single-frame exposure capabilities (exposing one film frame at the time), also provides the opportunity to *move* inert objects, to animate inanimate things at a speed selected by the camera operator. Shifting the placement of inert objects in front of a fixed movie camera, will make these objects move on the screen during projection. In this example, a camera exposes one frame of film while focusing on a wall with a portrait appearing at the left of the frame. Before the next frame is exposed, the portrait is hung a little farther to the right. For the next frame's exposure, the portrait is moved again, and successively this takes place for the next 21 frames. During projection the portrait's "movement" will mimic the action of our first example, the ball's going from point A to point B in one second.

The speed of the screen movement is controlled either by varying the number of exposed frames with an object in the same position in relation to the camera, or by varying the degree of displacement of the object between exposures. A combination of both actions will add flexibility to the animation technique and provide the resource to create a world in which objects appear to move with a life of their own. In this animated world, an apple will climb a stairway, a step at a time, where each step is recorded on the film, or it may climb up with high jumps that skip as many steps as exposed frames are skipped on the film. This simple technique has made possible the representation on film of extraordinarily complex actions. The gamut of animated films spans from the sequences of Walt Disney's cartoons through learning films that help us visualize actions of things that our eyes cannot detect, to recent epics such as *Star Wars*.

Film Transport

Film exposure and projection are achieved with the help of a revolving shutter and a pull-down claw. The sequence that obtains exact intermittent exposure or projection is as follows: the film is brought into the gate by the pull-down claw; the shutter opens allowing light to pass through the gate aperture; the shutter closes stopping the light; the pull-down claw engages the film sprocket holes and pulls down the film to bring the next frame into the gate; the shutter opens once again; and the cycle is repeated on and on 24 times per second. The shutter's opening and closing eliminates seeing the film movement as the film goes pass the gate aperture. The pull-down claw moves the film into position, lets it rest for the required exposure or projection time, and then pulls down the exact amount of film corresponding to the next frame (*see* Figure 9-15).

Figure 9-15. Shutter and Pull-down Mechanism of a Movie Projector

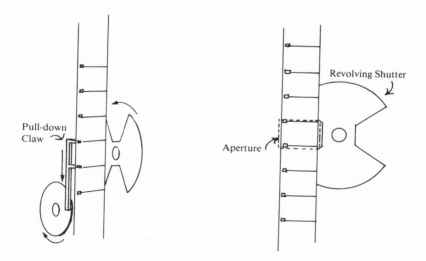

The black out occurring between frames due to its repetition rate (film running speed) does not show a "flicker" of the image; 24 frames per second adequately eliminates the problem. But earlier silent films run at a speed of 16 frames per second, and flicker is noticeable at this lower speed. The higher film speed is also useful for sound reproduction. We learned in chapter 7 that higher tape speed reproduces sound of better quality. This principle applies here as well.

Film Gauges

16mm and Super 8 are the film gauges of motion pictures most commonly employed in the audiovisual instruction field. Obviusly, 16mm film is twice as wide as Super 8 film. The term Super 8 differentiates between this newer format and earlier 8mm film, which had a frame of a different size. The position of the sprocket holes was changed, and this allowed for a bigger or "super" frame on film stock of the same width (*see* Figure 9-16).

Figure 9-16. Film Gauges

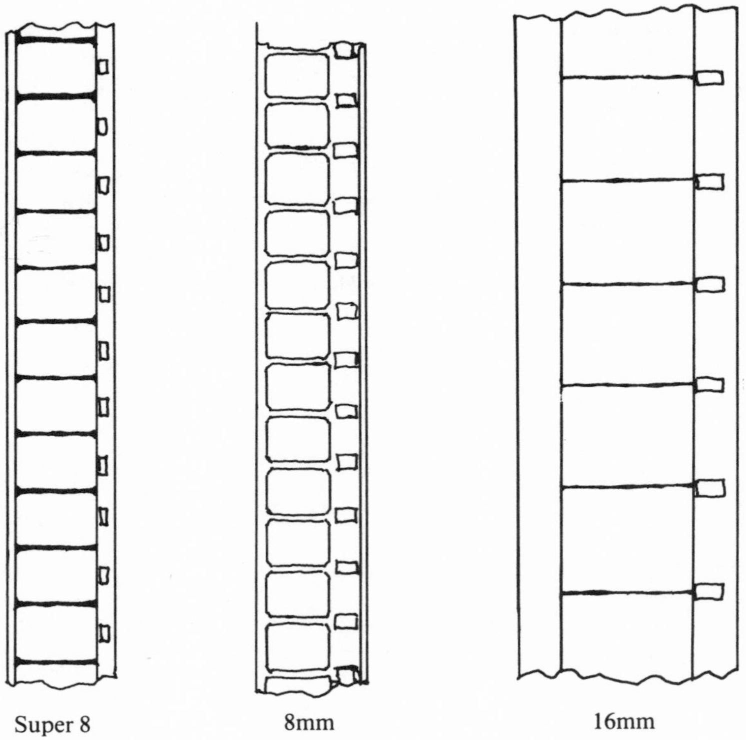

| Super 8 | 8mm | 16mm |

16mm Film

One foot of 16mm film has 40 frames on it. Close to one edge of the film are perforations used to transport the film through the camera or the projector. These perforations appear at the frame line, the line dividing adjacent frames. The area next to the other edge is reserved for the sound information. As previously indicated, the running speed of 16mm sound film is 24 frames per second and silent features run at 16 frames per second. Movie films are loaded on open reels usually at lengths of 800 or 1,200 feet for classroom films, although other sizes are also common, especially for 16mm copies of theatrical features. Figure 9-17 gives a sample of the running time of 16mm films of different lengths.

Figure 9-17. Chart of 16mm Film Running Time

Length	Sound	Silent
100 ft.	2 min. 47 sec.	3 min. 42 sec.
200 ft.	5 min. 33 sec.	7 min. 24 sec.
400 ft.	11 min. 7 sec.	14 min. 49 sec.
600 ft.	16 min. 40 sec.	22 min. 13 sec.
800 ft.	22 min. 13 sec.	29 min. 38 sec.
1,000 ft.	27 min. 47 sec.	37 min. 2 sec.

Super 8

One foot of Super 8 film has 72 frames on it. The sound area appears near the edge opposite the perforation side just as in the 16mm format. However, the perforations are lined up with the center of the film frame in this format. The running speed of Super 8 sound film is also 24 frames per second, but the silent speed is 18 frames per second. Super 8 films can be found in open reels or in continuous loop cartridges. The most commonly available cartridge contains 50 feet of film, although neither the length nor the physical form of cartridges is standardized. You must look for the projector that matches the cartridge and vice versa. Super 8 is still considered an amateur format and is mostly found in silent loops for subject contents that are acceptable for "single concept loops." A single concept loop consists of a sequence that demonstrates or describes a unit, a part of a larger whole. For example, it will explain one simple concept such as how to swing a bat rather than explaining the rules of baseball. These loops are better suited to small group or individualized instruction than to large audience projection. Super 8 film length is shorter than 16mm; Figure 9-18 gives a sample of running times.

Figure 9-18. Chart of Super 8 Film Running Time

Length	Sound	Silent
25 ft.	1 min. 15 sec.	1 min. 40 sec.
50 ft.	2 min. 30 sec.	3 min. 20 sec.
100 ft.	5 min. 0 sec.	6 min. 40 sec.
200 ft.	10 min. 0 sec.	13 min. 20 sec.
400 ft.	20 min. 0 sec.	26 min. 40 sec.

Magnetic and Optical Sound

Sound in motion pictures may be reproduced magnetically or optically. Magnetic film stock has a magnetic stripe applied to the film base. Audio information is recorded just as in an audiorecorder, which can be done either in the camera while filming or at the studio after the film has been processed. Optical sound is produced photographically. A light source exposes the film edge in correspondence with the sound information, creating an area with black, gray, and white values. In the projector, this photographic information is "read" by an "electric eye," a sensing device that converts this visual information into an electric current. This current is in turn amplified and heard over the loudspeakers of the sound system. Optical sound is the most common type of sound track found on commercially released 16mm films, though 16mm film projectors can be found with dual systems, optical and magnetic. On the other hand, Super 8 sound films are mainly produced with magnetic tracks. Optical sound is available, but it is still a rarity and is seldom encountered. Film with magnetic sound must also have a thinner magnetic strip next to the perforations for balance. Since a magnetic strip increases the thickness of the film at one edge, it is necessary to thicken the film at the other edge as well, so that the film is flat as it runs over the projection gate.

Projectors with Sound

Sound movie projectors have two reproducing systems, one for the visual image and the other for the programmed audio. The film leaves the supply reel placed before the projection gate, and usually goes through the sprocketed film drive. At the gate, the projection lamp casts the image on the screen as the shutter opens. The image is focused with the help of the projection lens (*see* Figure 9-19).

Figure 9-19. Diagram of a Movie Projector Film Path

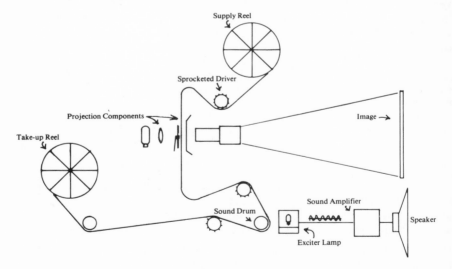

After the film leaves the gate, it goes over other sprocketed drivers which, incidentally, make up the main transport mechanism of the projector, before reaching the sound reproducing components. Here the film wraps around a "sound drum" with a shiny metallic surface. Opposite the drum there is an "exciter" lamp that, with the help of a small fixed lens, illuminates the sound area of the film. The light is mirrored back to a sensing device, which feeds the information to the amplifier. A take-up reel collects the film at the end of its run through the projector. In the case of a magnetic projector the sound components are replaced by an audio playback head similar to the ones found in sound reproducing equipment.

Looking at Figure 9-19, or at a projector, we can appreciate that there is a certain distance from the gate aperture to the sound drum. This distance is exactly 26 film frames for 16mm film and 18 frames for Super 8. This means that when the beginning picture frame appears in the projection gate, the accompanying sound should be heard over the loudspeaker. The sound information must be ahead of the picture on the film. It is important to keep this advance correct at all times to achieve "lip synchronization." The term indicates that if the image is a person speaking, the movement of the lips should correspond to the speech being heard. Projectors are threaded with a film loop before and after the gate. These loops, especially the one between the gate and the sound drum should be kept of the size prescribed by the manufacturer of the projector. The size of this loop will ensure that the sound and the picture will be in sync.

Film Splicing

The "misplacement" of the sound and picture is also apparent when viewing a torn film; the sound will not match the image. In splicing a torn film, you should try to keep all frames, which may not always be possible. Repairing a torn film requires a film splicer (*see* Figure 9-20), film cement, a cement brush, a dusting brush, and a pair of gloves.

Figure 9-20. Film Splicer

16mm film is spliced along the frame line, and cement splices are not visible. Super 8 film is usually cut across a frame and the splice is more objectionable. Basically, all splicers will hold the film in place, aligning the ends by the sprocket holes. The emulsion at one end of the film is scratched until the film base is exposed. The film is brushed clean and cement is applied over the scratched area. The other end is placed over the cemented area and allowed to dry under pressure for a few seconds. Practice splicing on an old film. It is an easy job, but a good and strong splice requires practice and care. Remember these points:

- Learn to know your splicer.

- Do not apply more cement than is necessary. Use enough cement to secure a good bond but not so much that it will run over the film spoiling frames adjacent to the splice.

- Always use fresh cement. Old cement will not bond strongly and the film will come apart, usually during projection.

- Handle the film with gloves and work on a clean surface to prevent dirtying and scratching the film.

OVERHEAD PROJECTION

A very useful projector employed widely in classrooms is the overhead projector nicknamed the "electric blackboard." This basic projecting device is different from the projectors introduced so far. It can project images in a well-lighted environment. The visual can be handmade, and even made during projection. Due to its design, the images are placed for projection "right side up." It is a great companion for a live presentation in which the lecturer improves the visual information as the lecture progresses. Its peculiar name, overhead projector, stems from the shape and optical system that beams the image over the user's head as the user faces the audience.

The Projector

The source of illumination is found at the base of the projector, and the light is beamed upwards through a transparent stage usually 10x10 inches (*see* Figure 9-21). Below the stage, there is a "fresnel lens," which diffuses the light evenly. The beam of light is intercepted by a lens that focuses it on a mirror angled at 45% from the vertical. The image is then reflected onto the screen through a front lens. The visual materials are placed on the stage and the image is focused by moving the "head" up and down.

Overhead Transparencies

Overhead transparencies or cells, the materials employed for projection, are sheets of clear acetate. The visual information can be written or drawn on these acetate sheets with a grease pencil or a felt tip pen. If the acetate is to be used repeatedly, the ink should be water soluble, to allow for erasing the information with a damp cloth. Grease pencil marks can also be wiped off easily with a cloth. Permanent ink pens can be used for durable transparencies. The pencil writings will appear black on the screen (the grease pencil balcks out the light, casting a shadow). Color is added with transparent inks. If a large area needs to be colored, self-adhesive transparent color acetates can be obtained and easily shaped and applied to the cell.

Figure 9-21. Diagram of an Overhead Projector

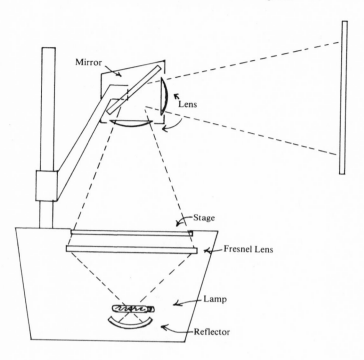

A transparency can be manipulated to show only a portion at one time. Since only transparent materials can be projected, by placing a sheet of paper on the stage, the transparency can be uncovered gradually. Using this projector, the outline of solid objects can be cast on the screen, since it is possible to place a three-dimensional object on the projector's stage. The lecturer facing the class can write on an acetate sheet placed on the screen, without the messy use of blackboard chalk and without turning his or her back to the audience. An acetate cell used as an "electric chalkboard" surface should be at least .5 mil thick for easy handling. Thinner sheets will not lie flat on the stage but will curl because of the projector's heat. Cardboard frames can be used to protect the transparencies and to provide easier handling. The cells can be attached to the frame by using pieces of masking tape. Two or more acetate sheets can be projected simultaneously. Used as the pages of a book, one cell can be superimposed onto another to add more information to a basic cell. This technique is described as using "overlays."

The Thermographic Copier

There are equipment and processes that make permanent black-and-white and color transparencies using photocopying principles. Xerox, 3-M, Agfa-Gevaert, and Arkwright are just a few companies that manufacture transparencies suitable for various processes. Here we will comment on the thermographic copier mentioned in chapter 5. A thermocopier uses a source of infrared light to expose a chemically treated film. This type of film is layed on top of the opaque original to be reproduced as a transparency (*see* Figure 9-22). Both are fed into the thermocopier, and are returned by it in a few seconds. The film acetate will have on it a burned-in image of the original material. There is one requirement for this method to work adequately. The original must have been drawn or printed with a carbon-based material. Graphite pencil, most printing inks, and typewriter ribbons have carbon on them and will work efficiently. If your original cannot be duplicated by this method, try making a photocopy to duplicate from. Photocopies are carbon-based.

Figure 9-22. Thermographic Copier

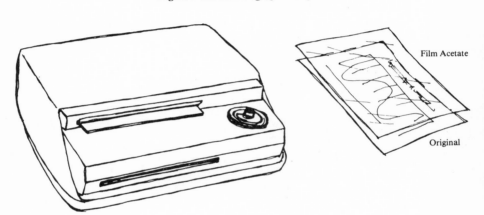

Film Acetate

Original

There are several types of these thermal acetates. Some will reproduce the information with a colored or black line on a clear background; others will duplicate an original from a book page without your having to tear off the page to feed into the copier. Some of these top loading machines may have intermediary steps that require a different, thinner sheet to be exposed first, and then, from this secondary or intermediate sheet, the actual overhead transparency is made. Thermal copiers will make opaque copies also. Mimeograph and ditto stencils can be produced by this copier. This is advantageous because the lecturer's notes can be handed out as copies of the projected materials; from the same original art work, transparencies can be obtained for projection together with hard copies for distribution.

Picture Transfer or Color Lift

Full color transparencies from magazine illustrations can be created by a simple and inexpensive method. The original illustrations will be destroyed and only images printed on clay coated paper can be "lifted," but the advantages of the process far outweigh its disadvantages. A full color illustration is generally printed on a clay coated paper to improve the adhesiveness of printing inks. That is to say, the paper serves as the base for a coat of clay that will be inked during printing. The picture transfer or color lift will remove the color inks and leave them attached to a clear acetate sheet to provide us with a color transparency for projection.

Although this method is simple, it requires careful handling of the materials. It was mentioned that the original illustration will be destroyed; therefore the beginner should practice prior to committing a wanted figure to this method. First, we must find out if the printing was done on clay coated paper, otherwise the image will not be lifted. With a wet finger, rub a corner of the illustration. A white, chalky residue left on the finger will indicate that the illustration can be transferred to acetate. Next, a piece of clear acetate must be bonded to the illustration. There are several materials and methods that can be employed in this step. A cold laminator using a pressure-sensitive, adhesive-coated laminating film, or heat laminators and dry mounting presses using laminating films are the best options. (Refer to chapter 10, "Dry Mounting.") The prescribed procedures for using these equipment and materials should be followed, making sure no dirt spots or air pockets are left on the illustration's surface. In using dry mounting presses it is suggested that a set of metal plates be employed for optimal results. The materials can be sandwiched between two cookie sheets or photographic plates to provide even pressure and good heat transfer. Clear contact vinyl will also provide a good bond and is readily available in hardware stores or art supply houses.

You may also create an adequate lamination for this purpose by scratching one side of a heavy-gauge acetate with fine-grain steel wool or sand paper. Remove all dust from the acetate and cover it with an even coat of rubber cement. Make certain that the cement flows easily and leaves no lumps or dirt. You should also apply a coat of rubber cement to the illustration to be lifted. To ensure a good bond, brush these two surfaces with strokes running at 90% angles from each other. For example, if you brushed the acetate with horizontal strokes, brush the illustration with vertical strokes or vice versa. Let both items dry thoroughly and bond them. (Refer to chapter 10, "Wet Mounting.")

To minimize the formation of air pockets when using this method or when employing clear contact paper, hold the acetate over the illustration folded in a U shape. Lower the acetate carefully, making the bottom of the U come in contact with the center of the illustration first. Afterwards, flatten the acetate rubbing from the center out so the air will escape through the sides. With the illustration on a hard, even surface, rub the materials with the aid of a flat blunt instrument, such as a burnisher or the back of a single-edged razor blade, to ensure a good, even bond. Remember always to move from the center toward the borders to release the air that may be trapped in the sandwich.

Once the illustration has been attached to the acetate by any of the methods described, dip the materials in a flat container of water. Adding a little mild liquid detergent or bleach added to the water will decrease the soaking necessary to remove the paper and clay from the ink. If dry mounting or heat laminating film was used, hot water will work faster. While the materials are in the water, place light weights on them to prevent thin acetate from curling. A pair of scissors or some clean pebbles may be enough for this purpose.

The length of the soaking necessary will vary depending on the method or the thickness of the paper base. After the paper has been peeled off, under running water and with the aid of a small wad of cotton or soft tissue, rub off the clay that still clings to the ink. During this step, observe great care not to press your fingers on the adhesive surface, because the ink will come off and the surface can be marred easily. This care is especially necessary if the rubber cement or clear contact vinyl method has been employed. Once the washing is completed, hang the transparency to dry. Thereafter, a coat of clear plastic spray should be applied to the dull side of the transparency to help protect the ink and increase the transparent quality of the acetate. A test of the plastic spray should be made when rubber cement or clear contact vinyl were employed. Some plastic sprays may dissolve the adhesive ruining the work.

A clear acetate sheet can be used to back the transparency and protect the delicate side. Transfers made using laminating or dry mounting methods can be better protected by applying a second sheet or film, which should be of a thicker gauge to minimize work. The process is now completed, although the transparency can be attached to a cardboard frame for easier and better handling.

It is also worth noting that when a magazine page has wanted illustrations printed back to back, laminating both sides of the page simultaneously will allow the capture of both images. Then by using a razor balde, split one corner and pull the two sheets of film apart carefully. Each inked image will adhere to the respective acetate sheet, and they can be further handled as explained above.

OPAQUE PROJECTORS

We close this chapter with a projector that, although it may not have the appeal and variety of applications of other projection devices, still has a singular advantage: it projects *opaque* materials, such as photographs, book pages, study prints, documents, and even small three-dimensional objects. This piece of equipment projects opaque material placed on the stage with the aid of a mirror and uses a lens to focus the image (*see* Figure 9-23). The stage is "opened" by lifting the upper part of the projector with a lever. It is usually a large bulky unit that requires a large source of light that, in turn, generates quite a bit of heat. A large, glass condenser must be used to protect the materials, if they are going to be shown for a long period of time. Unlike the overhead projector, an opaque projector must be used in a dark room for good visibility of the image. (Other, smaller, portable units of questionable quality can be rested directly on top of small opaque objects.)

Figure 9-23. Diagram of an Opaque Projector

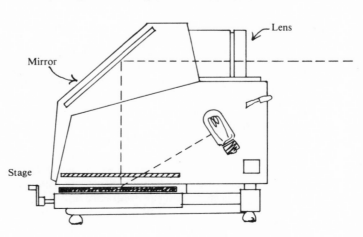

MULTIMEDIA INSTRUCTIONAL MATERIALS

Animated Motion. National Film Board of Canada, 1979. A series of instructional films directed by N. McLaren.

Animation: A Living Art Form. AIMS, 1971. 16mm film, 10 min., sound, color.

Facts about Films. International Films Bureau, 1975. 16mm film, 11 min., sound, color.

Opaque Projector, Its Purpose and Use. University of Iowa, 1958. 16mm film, 6 min., sound, B&W.

BASIC SOURCES

The Focal Encyclopedia of Film and Television Techniques. New York: Hastings House, 1969.

Kaufman, Jay M., and Lawrence Goldstein. *Into Film.* New York: Dutton, 1976.

Levitan, Eli L., *An Alphabetical Guide to Motion Picture, Television and Video Tape Production.* New York: McGraw Hill, 1970.

PART III
Audiovisual Media in the Communication Process

Old Circus Show

10 – THE SHOW

Our appreciation of a play, a film, or an art exhibit depends not only on its content and meaning but also on how it is presented to us. A Shakespearean monologue spoken with a marked foreign pronunciation, a movie seen with many interruptions, or a Rembrandt etching stuck to a wall with tape, will lose much of its appeal and its power to communicate. To communicate well, it is necessary to present information effectively and intelligibly. Information will be transferred more efficiently when the audience is favorably engaged by the delivery medium and not distracted or annoyed by it. Therefore, it is evident that in the communication process, the medium is just as important as the message itself, for the sender can only reach the receiver effectively via an efficient medium.

This is precisely the contribution that your technical role provides in the communication process. Your profession requires that you aid the delivery of information in an efficient manner. You are an important link in the chain leading to learning and much of the success in the transfer of information rests on how well you provide that link.

PRESENTING GRAPHICS

In presenting drawings, photographs, posters, or other types of displays, as much care should be given to the aesthetic appeal of the work as to the handling and preservation of it. A good drawing displayed against an appropriate background will attract attention. An appropriate and attractive mounting or setting enhances a work; a poorly presented art work will lose the appeal inherent in good graphics.

Selecting a Mount

In selecting a color matte or background board, a good rule is to choose a hue that is present on the graphic work. For example, a green pasture scene with spotted bunches of white flowers may be framed with a "whitish" border. But the choice is sometimes difficult; a single color dominates the work, a matte of similar hue will be overpowering and will detract from the work rather than enhance it. A gross example would be a white snow field on a white background where the art work and the background are fighting for dominance. Neutral hues and grays are a simple option in mounting black-and-white photographs of a good tonal variety. High contrast shots may work better on white or on deep black mattes.

In chapter 5, mention was made of allowing adequate margins on graphic work. This rule applies here as well. The size of the work will determine the width of the border on the matte. Be generous with margins. Do not crowd the display space. Allow for "breathing room" around the graphic work. The border will help to isolate the graphic from its surroundings, thereby enhancing its placement.

Cutting and Trimming

In cutting a matte or trimming a drawing, the choice of simple, sharp instruments, and a little practice will prevent ripping or tearing the edges of the materials and at times destroying the work. Work on a large, free surface. Cover the working surface with old newspapers to protect it from the cutting instruments, especially if you are using a drafting table for your working area. Cuts and holes on a drawing table will render it useless. You will never be able to use it for drawing again, because the knicks and holes left by the cutting instrument will serve as furrows where the pen or pencil will fall in later work. A soft disposable surface can replace the newspapers, but hard surfaces such as glass or metal are not recommended because, although they will protect the working surfaces, they dull the cutting instruments rapidly. One item recently introduced has the marvelous qualities of protecting the drawing surface, of being soft enough not to dull a blade, and of closing the slit or hole after the cutting or piercing to such an extent that the surface can serve as a good, smooth drawing area. This remarkable product comes in different sizes and in rolls that allow coverage of large working areas. It is a combination of rubber and plastic that "heals" itself after a cut. The sharpest cutting instrument will slide very easily on it, leaving no trace of the cut; the "nap" of this protecting product closes firmly after the blade has cut across its surface. Arttec® produces one brand of these cutting mats, and they can be found in graphic supplies stores.

A single-edged razor blade is a good instrument for trimming and cutting, although for better handling and control, an x-acto® or matte knife with a sharp blade is recommended. A straight edge, preferably of steel, is another necessary tool. A sharp blade will tend to cut, cooner or later, into a soft-edge ruler. Measure and mark your work before starting to cut. Remember the old saying of "measure twice, cut once." Lay your straight edge on the cutting mark, and slide your knife in one long, full-length cut. Do not stop and try to pickup the cut where you left off.

In cutting thick materials such as matte boards, cardboard, display panels, etc., do not attempt to cut in one stroke. Make one cut as described, and then go over it again and again until the work is done (*see* Figure 10-1).

Figure 10-1. Cutting Thick Materials

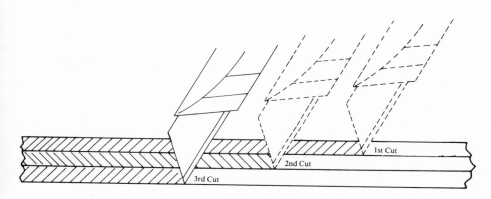

Trying to cut in one stroke will usually result in a crooked line. In trying to cut thick materials in one stroke, you will tend to increase the pressure of the cutting hand. Undue pressure will move the work or the ruler, usually as you draw your hand close to your body. Repeated passes of light or medium pressure will prevent damaging the work and cutting astray.

The angle of the cutting edge also contributes to making the cutting job easier. Hold the blade at approximately 30% from the horizontal plane (*see* Figure 10-2). In this position the blade will slide well. Keeping the blade too low will make the cut more difficult; too steep an angle will cause the blade to rip and tear the surface.

Figure 10-2. Angle of the Cutting Knife

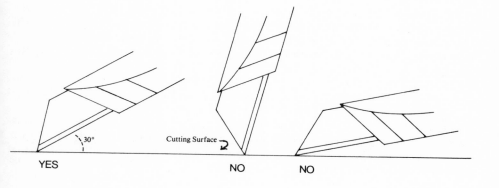

There are several types of paper-cutting devices that will aid in cutting. The most common type has a cutting surface where the material to be cut is placed. Attached to one edge of this surface is a long blade, which pivots, acting as a guillotine (*see* Figure 10-3). Unfortunately, these paper cutters are usually handled by too many persons and their blades tend to be out of alignment often. Be sure to check for knicks in the blade, usually caused by carelessly cutting papers with staples on them. Cutting several sheets simultaneously will also misalign the blade.

Figure 10-3. Paper Cutter

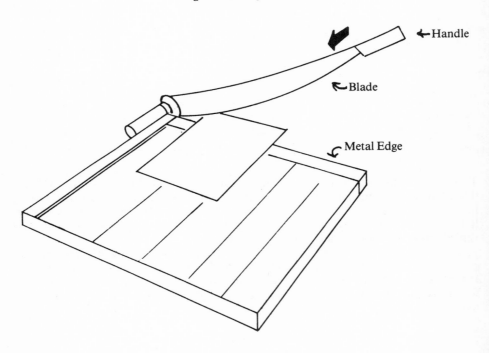

Before cutting the good material try the cutter on scrap paper. Always exert some pressure on the handle toward the board edge as you slice down. This will ensure that the cutting edge will press against the board edge in its downward movement, preventing the materials from slipping between the blade and the cutting edge.

Mounting

Once the art work is trimmed to size it may be necessary to mount it on a hard material. Fragile thin paper materials will benefit from being mounted on a sturdy backing. The backing will make displaying easier and will protect the materials from rough handling.

Wet Mounting

Mounting can be done by using common glues and "wet" adhesive compounds. This approach to mounting has several disadvantages. Using glues tends to be messy. Drippings can mar surfaces and spoil the work. At times, even great care will not eliminate these hazards. Wet gluing compounds dry slowly, and most will, in time, destroy paper products. Chemical components in glues act as acid, burning, corroding, and spotting the materials mounted.

A "wet glue" that is easy to handle and dries quickly is rubber cement. Although it will spot paper after a time, it is adequate for short-term mounting. It also has the advantage of providing a permanent bond or a removable bond, depending on the procedure used. For a nonpermanent mount, apply the cement to the backing and press the graphic in place while the cement is still wet. With this method, and with care, you will be able to peel the graphic off its mounting later on. If you prefer a permanent bond, apply rubber cement to both surfaces: the backing and the graphic, and let the cement dry thoroughly. Once the cement is dry, bring both materials together for a permanent bond. Remember that you cannot reposition the work once bonded, therefore be careful to mark the placement beforehand. You must also be careful in preventing air pockets from forming. Air pockets form easily, especially when handling large materials. It is good to start the contact at the center, smoothing the surface toward the edges. Observing this precaution allows the air to escape and thereby provides a good, flat contact area for the entire work.

A simple technique will aid this process well. In chapter 5 we mentioned release paper, a silicone-covered paper used for backing dry transferring materials. Release paper will not stick to rubber cement. Obtain two pieces of release paper large enough to cover the area to be glued. Once the coat of rubber cement is dried, place these pieces of release paper over the backing side by side covering the cemented area (*see* Figure 10-4A, page 226). Next, lay the graphic work over the release paper. Now slide to the side one of the pieces of release paper a little at a time while smoothing the graphic over the backing (*see* Figure 10-4B, page 226). Proceed always from the center out. Once one half is applied, repeat the same process with the other half. This simple technique will prevent blisters, a crooked mount, or broken graphics.

Among the wet glues, rubber cement is a handy product. You can cover the area without much worry about applying cement outside the area to be covered by the graphic material. You may want to spill over a little to ensure a good bond at the edges and at the corners. Apply it freely. Once the mounting is done and the cement is dry, the exposed cement can be removed easily with a rubber cement pickup. Go over the cement with this pickup and the work will be clean. You can even make your own pickup by collecting excess rubber cement as you rub it off with your fingers. You begin by making a ball that will get bigger as you remove excess rubber cement, forming your own rubber cement pickup. Although you may think rubber cement is an answer to your mounting problems, and indeed you will like it as you use it, remember that it is not everlasting and it will destroy the work after a time. Therefore, we must find a better way of mounting materials for preservation purposes.

Figure 10-4. Rubber Cement Mounting

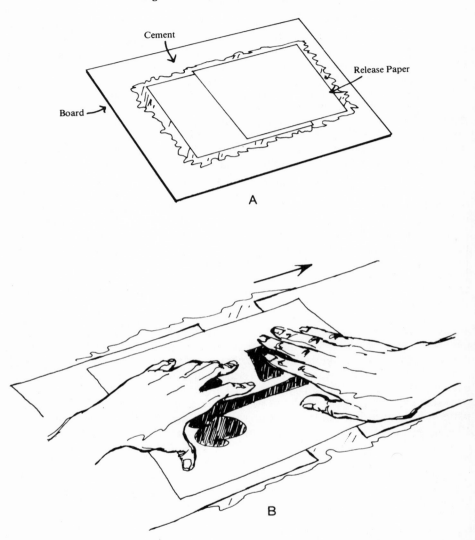

Cement

Release Paper

Board

A

B

Dry Mounting

Dry mounting has none of the disadvantages of wet glues. It is quick, clean, and will not deteriorate with time. Professional-looking displays can be achieved with this method and with a little practice and effort. Dry mounting also permits the mounting of photographs, paper, and cloth to cardboard, wood, Masonite, etc. Dry mounting utilizes as a bonding agent a tissue inert at room temperature. This tissue handles and looks like a sheet of waxed paper, but when exposed to heat it will fuse to provide a strong, permanent bond.

For dry mounting materials, in addition to the mounting tissue, a dry mounting press is required (*see* Figure 10-5). There are several sizes and types of presses. Some have thermostats to prevent the materials from scorching under excessive heat, and they may also have timing devices to control the heating cycle. But in essence, all presses have a flat, fixed bed, and a floating platen. The platen is raised by a lever-like handle, to allow the materials to be placed on the press. The same handle closes and locks the press for mounting.

Figure 10-5. Mounting Press

The mounting tissue, when inserted into the press and sandwiched between the work to be mounted and its backing, will bond these materials firmly. The dry mounting steps are as follows:

1) The material to be mounted, as well as the backing, must be dried to prevent moisture from escaping during mounting, which will cause blisters. Paper and cloth products will absorb moisture from the environment. Once on the heat press, this moisture will escape in the form of steam and will be caught between the product and the tissue forming an air pocket and preventing the bond (*see* Figure 10-6A, page 228). To prevent these blisters from forming, the materials to be used are placed in the press for a few seconds wrapped in brown wrapping paper. This paper will help absorb the moisture of the materials to be mounted.

Figure 10-6. Steps in Dry Mounting

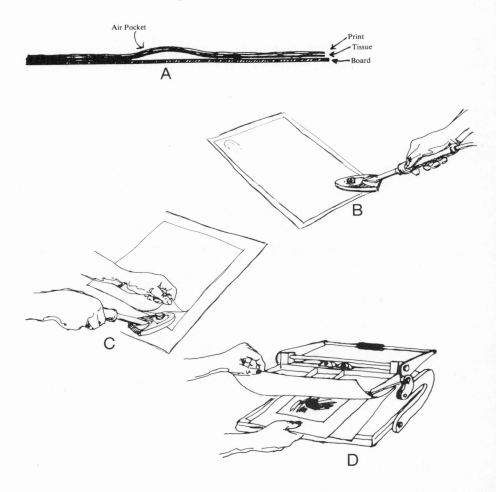

2) After the materials are dried, with the help of a tacking iron, a piece of tissue is tacked to the underside of the material to be mounted at one or two corners. A tacking iron is a hand-held instrument with a heat pad at one end as shown in Figure 10-6B. Use a piece of release paper between the iron and the tissue to prevent the tissue from sticking to the hot pad, and later ruining a clean surface by dirtying it with melted tissue.

3) Excess tissue is now trimmed, leaving only the back of the graphic work covered.

4) Now center the work on the board and tack the two remaining corners to it. This step will ensure that the work will not move when it is inserted in the press (*see* Figure 10-6C). (If a tacking iron is not available, a corner of the press platen can be used for

this light tacking. But always remember to use a piece of release paper over the mounting tissue.)

5) The materials are then inserted in a folder made of release paper. Previous careless mounting may have left melted tissue on the press platen. Molten tisue can ooze off the edges of materials slightly, and deposits of it can accumulate on the press. Release paper will protect the work while in the press.

6) Now the materials in the folder are placed in the press, and the press is closed. After the required time, the press is opened and the work is completed (*see* Figure 10-6D).

A good variety of dry mounting tissues are available with different characteristics and purposes. "Permanent" tissue provides a fix bond. "Removable" tissue will fuse when placed in the press for the first time; inserting the work again in the press on a later date allows for the materials to be removed from their mounts. Removable tissue also fuses at lower temperature, making it ideal for mounting delicate materials such as water colors, color inks, color photographs, etc., that may be affected by high heat. Cloth backing material is a thin cloth with adhesive on one side, making it ideally applicable for repairing maps, for backing materials to be rolled for storage, for hinging two or more flat pieces, for reinforcing plyable materials, etc. Laminating film can be applied over a surface to protect it. You can write on laminating film with a grease pencil or ink pen as explained in chapter 8. Lamination will protect materials from stains, moisture, and wear. Laminating can be done with a matte or a gloss finish. A matte finish will eliminate glare, and gloss finish will enhance colors. Thin materials to be laminated only and not mounted, should be laminated on both sides. This will prevent curling due to the surface tension of the lamination produced by slight shrinkage of the film in fusing.

Dry mounting materials fuse at different temperatures, and the length of time on the press differs in accordance with the thickness and type of materials to be mounted. It is best to follow the manufacturer's recommendation for each type of material. These materials are obtainable in sheets of different sizes and in rolls of varying length and width. The variety of available materials, and the simplicity of the process make dry mounting a desirable technique that affords good and professional results effortlessly.

SETTING UP FOR PROJECTION

Today, a great part of the work in resource centers deals with projection media. This work is most frequently done in a location away from the learning center; therefore, it requires preparation and forethought.

Scouting the Location

In answering a request for projection, first find out about the room conditions. Scout the location ahead of time. Make sure that the windows have shutters or blinds, if a film is to be shown or an opaque projector is to be used. Remember that you need a darkened room to see projected materials adequately.

Power Requirements

Electrical outlets should be accessible. They should also be near the projector's position. If they are not you will need to bring an extension cord. There are several types of outlets still around, but most equipment now has a parallel U-ground pin configuration (*see* Figure 10-7).

Figure 10-7. AC Power Plugs

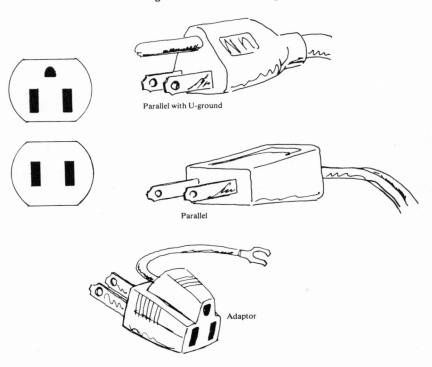

Parallel with U-ground

Parallel

Adaptor

Many buildings still may have a simple parallel outlet. An adaptor will be needed in this case. Electric current (amperage) available in each outlet is, for the most part, sufficient for one projector. Standard outlet supply is rated at 15 amps. A simple formula will help you find out how many units you can operate safely in one room:

$$\frac{\text{Watts}}{\text{Volts}} = \text{amps}$$

Dividing the power consumption of the unit by the volts of the supply will indicate the amperage required. For example, a projector lamp rated at 1,000 watts (our electrical supply is always 120 volts) will require 8.3 amps.

$$\frac{1,000 \text{ watts}}{120 \text{ volts}} = 8.3 \text{ amps}$$

Therefore, two projectors, each with a 1,000-watt lamp, will need 16.6 amps (8.3 amps x 2 = 16.6 amps). Since available supply provides only 15 amps, this current overload would, in most cases, blow a fuse. This indication about amperage consumption is here to alert you. In most cases, all you will use is one movie or opaque projector at a time. These units, aside from lighting fixtures used for photographic work, are the heaviest users of energy you will employ. Most other electrical units encountered in your work are low consumers of energy and do not require heavy amperage.

Screens

If the room walls are painted with an off-white color and one wall is free of objects, you may use it as a projection screen. If not, you will need to bring a screen.

There are basically three types of screen surfaces: matte white, glass beaded, and silver lenticular. Each is best adapted to different viewing conditions. The matte white screen is used where the projection light source is strong, such as in overhead projection. This screen will diffuse light evenly over a large surface area. Its smooth, non-gloss surface will reproduce a picture with good detail. Glass beaded screen surfaces are several times brighter than matte white. Glass beaded screens have excellent picture sharpness and good color rendition. The room should be darkened for good reproduction. Glass beaded screens have a narrow viewing angle. The image is brighter and sharper when viewed head-on than when viewed at an angle. A lenticular screen is designed to control horizontal light reflections. They can be used in partially darkened environments and are extremely sharp and brilliant.

Seating

Now, let's turn our attention to the seating arrangement and screen placement. Most rooms will have windows, and even with shutters or blinds, there may be some light entering the room. In most classroom situations this condition may be beneficial because it will allow the students to see sufficiently to take notes. From a projection standpoint it is not ideal. If possible, place the screen back to the windows. This will prevent light from "spilling" over the screen and "washing out" the projected image. The screen should be placed where it can be well seen. Arrange the chairs to provide a clear line of vision, and leave a clear path for the projector's beam (see Figure 10-8, page 232). A better choice than leaving a clear path for the projector beam is to use a high projection stand. A projector set up 4 feet high will not interfere with a person of average height seated on a standard seat.

Image Size

The audience and room size will determine the screen size and the projection lens needed for comfortable viewing. Screen size depends not only on visibility but on the effect desired on the viewer. A three-foot wide screen is just as visible as a Cinemascopic screen when viewed from 10 feet, but the effect isn't quite the same. The visual impact of a large screen is independent of readability. In any case, for good viewing the farthest seat from the screen should not be more than five or six times the screen width. Sitting too close to a large screen is also

Figure 10-8. Suggested Seating Layouts

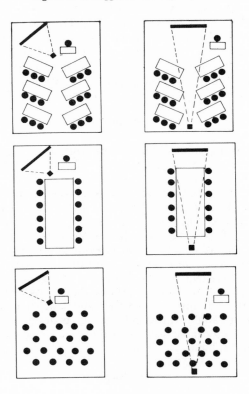

uncomfortable. The screen should be removed from the front row. Once the screen is obtained, you must fill it with a projected image. This can be simply accomplished by moving the projector closer or farther away from the screen to vary the projected image size. Getting the projector closer to the screen will reduce the image; moving the projector away from the screen will increase the image size. In most situations you will encounter, this method will provide an adequate solution. However, in some cases you may wish to have a stationary projector location and the screen at a fixed distance from it. This would be the case in an auditorium that may have a built-in screen and a projection booth to keep the projector sound from interfering with the showing. The chart in Figure 10-9 will help you identify which lens will be needed depending on these variables: picture width and screen-to-projector distance. A line from the left margin of the chart (screen distance) will intersect with a line from the top (picture width/16mm projector) or bottom margins (picture width Super 8 projector) at a point on one of the diagonals indicating the lens size required. This chart also allows you to know the type of screen width you will need for a given lens at a known distance. For example, in a classroom 20 feet long using a 16mm projector with a 1½-inch lens, you will need a screen five feet wide. You arrive at this solution by drawing a horizontal line from the left margin to the diagonal 1½-inch lens line, and from this intersection a vertical line up to the upper margin, which indicates a five-foot picture width. It follows that using this chart for a given screen width and lens size you will know the distance needed from the screen to the projector.

Figure 10-9. Lens Selection Chart

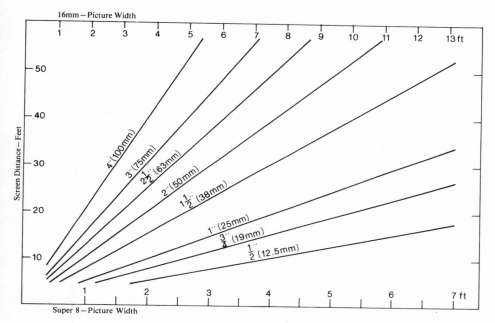

Television Viewing

In television viewing it is recommended not to sit too close to the set. A minimum viewing distance is five or six feet from a large screen size. Television screens are measured diagonally in inches. A ratio of one inch to one foot indicates the maximum viewing distance, i.e., the last seating row in front of a 19-inch television set should not be further than 19 feet from the television set. You should also be aware that a viewer sitting too far off the center of the television viewing axis will see a distorted picture. The angle of viewing of a television screen should not be more than 45° from the 0° axis. Figure 10-10 (page 234) shows the seating requirements for adequate viewing.

Figure 10-10. Television Viewing

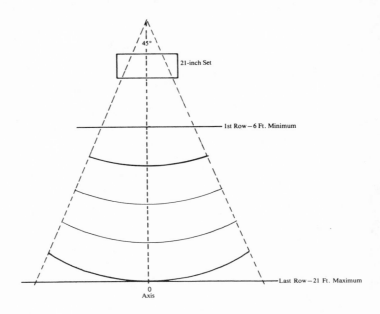

Keystoning

Viewing a flat surface from an oblique angle will result in a "keystone" appearance of the image. Keystone is the distortion of a rectangular or a square area that shows one side shorter than its opposite, parallel (*see* Figure 10-11). Keystoning is also present when an image is projected on a screen surface that is not parallel to the projector's film aperture plane. The screen and the projector must be aligned to eliminate keystoning.

The example of the overhead projector in Figure 10-11 should illustrate keystoning distortion clearly. To solve this problem the screen is tilted at the top toward the projector, presenting a plane parallel to the lens head. This same principle applies when keystoning is encountered with any other type of projection equipment.

GETTING READY

In preparing for a recording or a presentation, check every unit to be used. Be certain that everything is operating well. Testing beforehand will eliminate embarrassment, and will provide the needed support service without inconvenience or delays.

A Reminder List

- Make sure you have a take-up reel for a film projector or an audio or videorecorder that is of the same size as the supply reel.

Figure 10-11. Keystoning

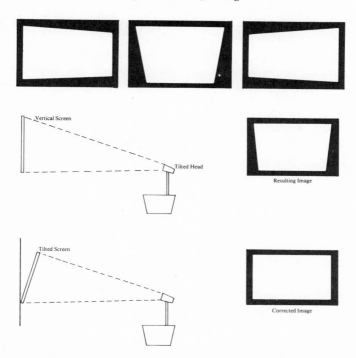

- Check to see that you have the correct power cords and that they are in good condition. In many detachable power cords, the inner conductors tend to break at the plug or jacks after repeated use.

- Always take a spare projection bulb. After many hours of use, these bulbs will burn out, usually when the unit is switched on.

- Clean the projection gate and lenses in the case of projectors, and clean video and audiorecording or playback heads in magnetic recording or playback equipment.

- Bring the correct connecting cables: projector to extension speakers, monitor to playback, microphone to mixer or amplifier, etc.

- Try microphone and audio extension cables for continuity. Broken leads will not transmit in case of complete defect or will introduce "buzzing" on audio signals when partial breakdown occurs.

- For audiorecording sessions, do not forget microphone stands to hold the microphones.

- Bring your recording tape and an extra supply of it. Many activities may be estimated to last a predetermined time, but often they may be prolonged unexpectedly. Be prepared, and you will capture the entire program.

- Set up equipment ahead of time. Do not arrive after the audience has gathered or the speaker has begun. Remember that your job should be as unobtrusive as possible. You are there to help and not to distract or create confusion.

- Place long runs of cable away from traffic areas. Run cables along walls or under chairs and tables. If the cables stretch across walkways, tape them to the floor. Loose cables are a hazard to people. In public events, people are not attentive to these trappings and they may trip over cables and equipment. Again, remember that it is your job to pay attention to these details and to prevent accidents connected with your presence.

- Be neat in stretching cables and setting up equipment. Do not leave empty film or tape containers around. Place equipment covers and other accessories out of reach. They are not only probable hazards, they may get broken or lost, and an untidy setup reflects on your professional performance.

- Test microphone recording or playback volume well in advance. When the speaker is ready or the audience is waiting, the presentation should begin. That is not the moment for you to do the testing. Again, avoid distracting attention from the event.

- In film projection or video playback, thread the film or tape, and find the beginning of the program well in advance. Stop the equipment on the right spot, and wait for the start. When the show starts, useful information should be displayed on the screen. The audience is not there to see a few feet of leading materials used for technical reasons. Film leaders have a countdown sequence and you should stop at number 2. This brief interval is sufficient for the projector to come to constant speed before you turn the lamp on. In a commercial videotape usually a video test signal is accompanied by a sound level signal and at times also a countdown.

- Handle all materials similarly; be ready to begin when the lights go out.

- Project visuals to ensure that they are right side up. In the case of slides, look at the entire set. A single reversed slide can ruin an otherwise good presentation. Running the set once will also check for slide damage that may prevent them from entering the projector's gate.

In summary, the following activities should be completed in getting ready for a program:

1) Check all components well in advance.
2) Arrive ahead of time.
3) Set and test all equipment prior to show time.
4) Be as unobtrusive as possible.

With advanced planning and thought, the show can be a very successful experience for the audience and yourself.

MULTIMEDIA INSTRUCTIONAL MATERIALS

Dry Mounting Instructional Materials: Basic Techniques. University of Iowa, 1965. 16mm film, 5 min., sound, color.

Dry Mounting Instructional Materials: Creative Applications. University of Iowa, 1969. 16mm film, 7 min., sound, color.

Dry Mounting Instructional Materials: Laminating and Lifting. University of Iowa, 1965. 16mm film, 6 min., sound, color.

Facts about Projection. International Film Bureau, 1976. 16mm film, 16 min., sound, color.

Rubber Cement Mounting. McGraw Hill Films, 1965. Single Concept 8mm coop, silent, B&W.

GLOSSARY

Accession number — A number given to an item indicating the order in which it was received and added to the collection.

Accompanying material — Dependent materials such as teaching and learning guides provided with an item as additional aids.

Acetate cell — *See* Cell.

Acquisitions department — The department responsible for the ordering and procurement of materials through either purchase, exchange, or gifts.

Added entry — Any access point other than the main entry made in the catalog.

AECT — The Association for Educational Communications and Technology, a national organization for specialists and others in the field of audiovisual instruction and technology.

ALA — The American Library Association, the national organization for librarians and others interested in libraries and library-related matters.

Ampere — *See* Current.

Amplifier — A device used to increase or reinforce the strength of an electronic signal.

Animation — The technique of filming a number of slightly different drawings, or objects of slightly different placement in relation to the camera, so as to create the illusion of movement.

Annotation — An evaluation, description or explanation included with an entry in a list, bibliography, or catalog.

Annual — A serial publication issued consistently once a year.

Aperture — The opening in a lens, a camera, or a projector through which light passes.

Art print — A reproduction of a two-dimensional work of art.

Audiocylinder — *See* Audiorecording.

Audiodisc — *See* Audiorecording.

Audiorecording — A recording of sound in either disc, cylinder, roll, wire or magnetic tape format.

Audioroll — *See* Audiorecording.

Audiotape — *See* Audiorecording.

Audiovisual materials — Materials requiring the use of special equipment to be seen or heard. *See also* Media.

Audiowire — *See* Audiorecording.

Azimuth — The angular relationship of the head gap to the tape path on magnetic recording or reproducing equipment.

Balance — 1) The harmonious proportion or arrangement of parts in a work of art; 2) the proper relationship between the level of two or more sound sources.

Balanced line — An electrical line with two conductors and a shield in which the signal lines are at the same potential in relation to the shield or ground.

Bass — The low frequencies in the audiospectrum.

Bleeding — 1) The running of a fluid medium; 2) to extend an illustration beyond the usual edge.

Burnisher — An instrument with a blunt, rounded point made of bone, plastic, or wood used for applying dry transfer materials.

Call number — Identification symbols consisting of numbers and/or letters which indicate the subject classification and the location of a work on the library or media center shelf.

Calligraphy — The art of freehand writing; penmanship.

Capacitor — An electronic component that stores energy. Also called condenser.

Capstan — A motor-driven metal rod that moves magnetic tape at a constant speed past the recording or playback head.

Card catalog — Entries arranged in drawers in a definite order giving the holdings of the learning center or library.

Carrel — A study desk or cubicle designed for individual study.

Cartridge — An enclosed container for film, microfilm, audiotape, or videotape in an endless loop format.

Cassette — An enclosed container for film or magnetic tape in reel-to-reel format.

Cell — A sheet of transparent acetate or celluloid used for projection. *See also* Transparency.

Chart — A sheet of information in graphic or tabular form.

Chroma — Color intensity.

Chromatic — Pertaining to color.

Classification — A systematic scheme for the arrangement of materials according to subject or form.

Collation — Description of a work which indicates the physical characteristics of an item.

Condenser — *See* Capacitor.

Contour — Line defining a form or figure; outline.

Contrast — The degree of intensity of light and dark, or of the chromatic values of objects.

Current — The flow of electric energy measured in amperes (abbr. amps).

Cycle — A complete oscillation of a vibrating or alternating object or force. Also referred as Hertz (abbr. Hz).

Decibel — A unit equal ten times to the logarithm of the ration of two powers. It is used in sound volume measurement.

Degausser — An instrument used to demagnetize recording heads and other metallic surfaces of magnetic recording or reproducing equipment.

Depth of field — The area in which all objects in a photograph or a film are in focus.

Diaphragm — 1) The adjustable opening formed by overlapping plates controlling the light passing through a camera lens — also called an iris; 2) the vibrating membrane of a microphone or a loudspeaker.

Discography — A list of recordings on a specific subject or by a particular artist.

Distortion — Unwanted change on a response signal as it is processed by electronic equipment or when it is transferred from one medium to another.

Distributor — Agency or company which generally has the legal rights of distribution and from which materials may be purchased, borrowed, or rented.

Dry mounting — A bonding process that uses heat and a fusing tissue to seal or mount materials.

Edition — The complete number of copies of a work produced from the same master and issued at one time or at intervals.

Exposure — Subjecting a photosensitive surface to light.

f-**stop** — The calibration of a lens indicating its apperture.

Fastmotion — The presentation in film of action that appears faster than normal.

Feedback — A squealing sound from a loudspeaker, produced by re-entry of the sound into the microphone.

Field — Half of a television frame.

Film — A transparent strip of celluloid or acetate with images or frames produced photographically. *See also* Transparency.

Filmloop — Film spliced in a loop for continuous playing without rewinding.

Filmstrip — Usually a 16mm or 35mm roll of film that presents a succession of images to be viewed frame by frame with or without sound.

Focal length — The distance from the optical centers of a lens and the film plane or focus point. The focal length of a lens determines the area of coverage.

Font — A full assortment of printing type of a particular face, or style, and size.

Frame — 1) A single image on a strip of picture film; 2) in television, a complete picture made up of two interlaced fields.

Frequency — Repetition rate of electrical impulses, usually expressed in cycles per second.

Gelatine — Gel for short, a colored material used for filtering light.

Gloss — High luster finish; shine.

Harmony — A state or order in the relationships of parts to the whole.

Hertz — *See* Cycle.

Hue — A particular shade or tint of a color.

Impedance — Resistance to electric current flow presented by certain physical characteristics of electronic circuits. It is measured in ohms.

Imprint — Part of the catalog entry which indicates the place of publication, the producer, publisher, distributor or sponsor, and the date of publication or distribution.

Iris — *See* Diaphragm.

Kelvin scale — The absolute scale of temperature. Kelvin degrees are used to indicate the color temperature of light sources.

Kinescope — A motion picture photographed from a television screen.

Kit — Two or more media which are subject related and intended for use as an instructional unit.

Lantern slide — A piece of glass with an image or words used for projection with the help of a lantern slide projector. *See also* Transparency.

Lithography — A printing method using a stone, in earlier days, and presently a metal plate, that bases its technique on the grease-repels-water principle.

Lithostone — A type of fine-textured limestone used in lithography printing.

Loop — A short length of film, videotape, or audiotape in a continuous loop designed for payback without rewinding.

Magnetic tape — A plastic ribbon coated with iron oxide used to record magnetic impulses from digital video or audio sources.

Matte — 1) A border or frame for a picture; 2) a dull, lusterless finish.

Media — Plural of medium. 1) A basic material used in artwork such as pencil, ink, charcoal, oil paint, watercolor, etc.; 2) a channel of communication or a technology employed to communicate, such as television, newspapers, books, movies, etc.

Medium — *See* Media.

Microfiche — A flat sheet of film containing microimages that can be used only with magnification.

Microfilm — Microimages on a roll of film that cannot be utilized without magnification.

Motion picture film — A transparent film, with or without a soundtrack, bearing a sequence of images that create the illusion of movement when projected in rapid succession.

Newton rings — A pattern of rings of distorted shapes that are produced when two surfaces are separated by a small distance and illuminated.

Nib — The point of a writing instrument or pen.

Offset printing — A method of transferring multiple copies of an image or a printed text; also known as picture transfer.

Ohm — Unit of measurement of resistance to an electric current on an electric circuit. It is represented by the Greek letter Omega (Ω).

Order department — *See* Acquisitions department.

Overhead transparency — An acetate cell or sheet bearing images or words that can be projected onto a screen with the help of an overhead projector. *See also* Transparency.

Picture — A two-dimensional work of art in either an art print, photograph, or study print format.

Picture transfer — *See* Offset Printing.

Pigment — Substance used for coloring.

Point — The unit measurement for printing type, equivalent to 1/72 of an inch.

Previewing — The screening of an item in order to select it for purchase or rental.

Quill — A pen made from a feather; a pen.

Release paper — A silicate sheet used to protect surfaces from glue or other adhesives.

Scanning — The process by which a television system reproduces an image electronically.

Shutter — A device used to block off light from the film in a camera.

Slide — A transparent positive image on a frame of projection film. *See also* Transparency.

Slow motion — The presentation in film of action that appears slower than normal.

Tacking iron -- An instrument used in dry mounting. It has a "heat pad" employed to melt dry mounting tissue or other materials.

Texture — The tactile characteristics or appearance of a surface.

Transparency — A piece of celluloid, acetate, or glass bearing an image that is viewed by projecting light through it. *See also* Overhead transparency, Slide, Lantern slide, Film.

Treble — The high frequencies in the audio spectrum.

Tripod — A three-legged camera mount.

Ultrafiche—Microimages on a flat sheet that are reduced 90 times the original size.

Unbalanced line—An electrical line with one conductor and a shielf or ground.

Vertical file—Clippings, pamphlets, pictures, and other such items filed in drawers for ready reference.

Video—The portion of a television signal that carries the visual information.

Videocassette—A videorecording on magnetic tape enclosed in a cassette.

Videodisc—A videorecording on a disc.

Videorecording—A recording designed for television playback.

Viewfinder—The camera component through which the image viewed by the lens can be seen.

Wet mounting—A process for bonding that uses glue or adhesives.

INDEX

AACR 2. *See Anglo-American Cataloguing Rules* (2nd ed.)
Academic center, 24-25
Access points, 84
Accession number systems, 75
Acoustics, 165
Added entry, 82, 84
Adhesive labels, 90
Adhesives, 225-26
AECT. *See* Association for Educational Communications and Technology
ALA. *See* American Library Association
ALA Rules for Filing Catalog Cards, 85-87
American Library Association, 46, 54, 56
American National Standards Institute (ANSI), 201
Ampex Corporation, 178, 185
Amplifiers (audio), 165-66
Amplifiers (video), 183-84
Anderson, Donald M., 128
Anglo-American Cataloguing Rules (2nd ed.), 82, 83
Animation, 206-207
ANSI. *See* American National Standards Institute
Anti-Newton-ring glass, 194
Armat, Thomas, 204
Art of Written Forms, 128
Arttec® cutting mat, 222
ASA number, 139
Associate specialist qualifications, 51-52
Association for Educational Communications and Technology, 22-23, 46, 54, 56
Asymmetrical composition, 114
Audio cables, 163-64
Audio response curve, 166
Audio spectrum, 157
Audiocassettes, 169-70
  See also Audiorecording tape; Audiorecordings
Audiodiscs
  labeling and storage, 92
  See also Audiorecordings
Audiorecording head, 167, 169-71
Audiorecording tape
  care of, 175
  format, 168-71
  speed, 168
  splicing, 175-77
Audiorecordings, 31, 156-77
  history, 156
  labeling and storage, 98
  monophonic, 172
  stereophonic, 172
Audiovisual center, 23
  corporate (chart), 41
  defined, 22, 23
  facilities, 55-59
  furniture, 60

organization and administration, 39-41
  charts, 40, 43
  personnel, 45-53
  production area, 56
  services, 42-44
  storage, 56
  See also Learning center
Audiovisual equipment, 33-37
Audiovisual Instruction, 63
Audiovisual Marketplace: A Multi-Media Guide, 69
Audiovisual materials. *See* Educational media
Audiovisual Materials, 66-67
Author number, 79
Available light photography, 139

Baird, J. L., 178
Balance (design), 113
Balanced transmission line, 163-64
Berliner, Emile, 156
Bidirectional microphone, 161
Bleeding (lettering), 117
Book catalog, 81
Booklist, 65
Braun, K. F., 178
Burnishing (dry transfer), 125

Call numbers, 72, 79-81
Camera, 136
Camera obscura, 136
Campbell Swinton, A. A., 178
Capacitor, 158
Capstan, 173-74
Captions, on visual displays, 115-16
Card catalog, 81
Cardioid microphone, 161, 162
Career lattices, 51, 53
Carousel projector, 193, 194
Carrels, 60
Cassette tapes. *See* Audiorecordings
Catalog, 68-69, 81-87
Catalog cards, 84-87
  arrangement of elements on, 82
  filing, 85-87
Cataloging, 71-87
Cathode ray rube, 179
Cells. *See* Transparencies
Charts, labeling and storage, 103-105
Cinematograph, 204
Circulation cards, 90-91
Clarity (design), 114
Classification schemes, 74-78
Clerk, qualifications, 51
Color, 113
Color film, 143
Color lift, 214-16
Colorimetry, 141-43